BARRON'S

NEW JERSEY

GRADE **6**

MATH TEST

Mary Elizabeth Platt, B.S., M.Ed.

About the Author

Mary Elizabeth Platt is a life-long resident of New Jersey. She earned her B.S. in Education from Chestnut Hill College, Philadelphia and her M.Ed. from Rutgers University, New Brunswick. Currently she is in her 11th year as a teacher in Hamilton Township Public Schools, Mercer County, NJ.

Acknowledgments

For my friends, family, and colleagues for their boundless encouragement, humor, and advice. Thank you to M.D. for the opportunity and support and C.C., P.G., M.S., and C.T. for their mathematical expertise. Finally, for my students, who teach me about being a good educator every day.

All inquiries should be addressed to:
Barron's Educational Series, Inc.
250 Wireless Boulevard
Hauppauge, NY 11788
www.barronseduc.com

ISBN: 978-1-4380-0723-6
Library of Congress Control Number 2015934505

Date of Manufacture: October 2015
Manufactured by: B11R11

Printed in the United States of America
9 8 7 6 5 4 3 2 1

10%
POST-CONSUMER
WASTE
Paper contains a minimum
of 10% post-consumer
waste (PCW). Paper used
in this book was derived
from certified, sustainable
forestlands.

Contents

IMPORTANT NOTE: Barron's has made every effort to ensure the content of this book is accurate as of press time, but the PARCC Assessments are constantly changing. Be sure to consult *https://www.parcconline.org/* for all the latest testing information. Regardless of the changes that may be announced after press time, this book will still provide a strong framework for sixth-grade students preparing for the assessment.

Introduction

CHAPTER 1

What Is the Common Core?

The idea of standards is not new; for many years, each state has had a list of criteria for what should be taught in each grade level. Recently, many states have joined together to create a single vision for what should be taught in our schools. This vision is referred to as Common Core. The Common Core has created a path for students where they are supposed to achieve mastery of specific skills at certain grade levels. The goal is that all students who graduate from high school are capable of success in an increasingly competitive global society. Developed in 2009, Common Core has been adopted by 43 states.

Each grade level has a set of standards made up of broad categories called domains. The domains for sixth grade are the Number System, Ratios and Proportions, Expressions and Equations, Geometry, and Statistics and Probability. These five domains comprise the five main chapters of this book. Each domain is separated into clusters, or categories, and then strands, which are the individual skills set. This creates the list of all the material that you need to master in sixth grade, and therefore creates the basis for the types of questions you will see on the PARCC test.

For details about the sixth-grade Common Core Math Standards, see Appendix A.

What Is the PARCC Test?

I am sure that you have heard the word PARCC many times. You have probably thought your teachers and parents were talking about a "PARK" test. PARCC is an acronym for an educational group: the Partnership for the Assessment for the Readiness of College and Careers. This is a really fancy way of saying they want to know if the students in our schools will, one day, be ready for the difficulty of college and the responsibility of a job. The goal is that all students work at a challenging level so that they can achieve their highest potential.

This group has put together a set of tests for each grade level to assess where students are in meeting grade-level standards. These standardized tests will be taken by every student in the states that are part of the PARCC. You have taken standardized tests before, and just like the versions you have taken in the past, the PARCC test will focus on your Mathematics and English Language Arts knowledge. This book will look specifically at the Mathematics portion of the test for sixth graders.

Computer Assessments

One of the biggest differences between the PARCC test and any other test you have taken before is that it will be given entirely on a computer. This will make many of the questions very different. You won't be able to make drawings the same way and will have to show your work and do your computations in other ways. Additionally, there will be a variety of tools available to you as part of the test.

Some tools will be available at the top of the screen. This includes a ruler, a protractor, and the answer eliminator (with which you can cross out wrong answers), and for some questions this is where the calculator is located. The calculator available to you will be a basic four function calculator that can only do a single step at a time. On previous tests, you have probably had access to calculators that did more advanced operations. If you click the outline of the person at the very top right corner, you can also use a magnifier and a focusing tool for your comfort.

There is also a highlight feature on most problems. Simply left-click and drag your cursor or touch pad (just as you would highlight any text). A pop-up box will appear, and you can select yellow, blue, or pink highlighting. Take some time and complete the online tutorial. Two different tutorials will help make your testing go smoothly and set you up for confidence. Make sure you visit these sites!

1. *http://epat-parcc.testnav.com/client/index.html#getitem/7443*

2. *http://epat-parcc.testnav.com/client/index.html#tests*

You will have to use the equation editors to do much of your work. This is how you type your answers into the system. There are two types of equation editors. The basic equation editor is for entering numerical responses and simple expressions and equations. In this program, you will not be able to type words or sentences. At other times, you will need to enter an extended response, which will allow for a combination of words, numbers, and equations. The second type is available when you are asked to explain, defend, or describe.

There are quite a few buttons available in the equation editors for the vast number of symbols you will need. The buttons are designed for all students in Grades 6–8, and not all of them will be needed for the sixth-grade test. The PARCC has set up a tutorial so that you can familiarize yourself with all of these buttons. This can be found at *http://epat-parcc.testnav.com/client/index.html#tests* or by searching PARCC Equation Editor Tutorial. Be sure you play with this tutorial before the test so that you are comfortable with how you input your answers. If you hold your cursor over the button without clicking it, a pop-up box will appear telling you the name of that sign. This is one instance where a little bit of practice can have a huge impact.

It is important to note that even though all parts of the test will be taken on a computer, many parts of the PARCC Assessment will be scored by a human scorer. We will look more at this when we examine the different types of questions.

Types of Questions

Type 1: Understanding Content and Fluency; Applying Mathematical Rules

This is your basic operation question. You may be asked to solve an equation, complete a table, use a drop-down menu to complete a statement, or drag and drop items into their correct placement.

The area of a rectangular patio is $5\frac{5}{8}$ square yards, and its length is $1\frac{1}{2}$ yards. What is the patio's width, in yards?

○ A. $3\frac{3}{4}$

○ B. $4\frac{1}{8}$

○ C. $7\frac{1}{8}$

○ D. $8\frac{7}{16}$

Type 2: Arguments, Reasons, and Critiques; Explaining Math

This is a question that is going to ask you to explain, describe, or defend. You may have to explain why your answer is correct, how you followed the steps, or why a given answer is correct/incorrect. This often means you should be able to put your mathematical thinking in words.

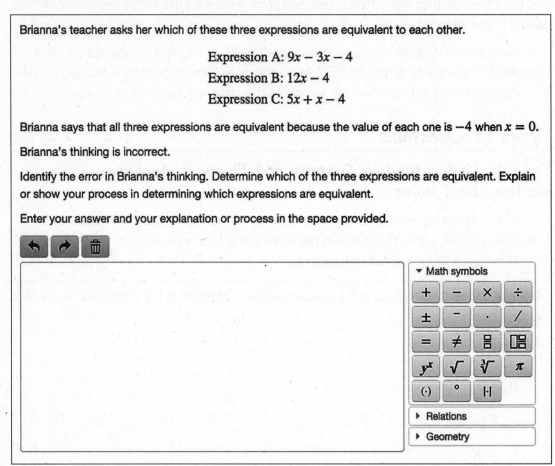

Brianna's teacher asks her which of these three expressions are equivalent to each other.

$$\text{Expression A: } 9x - 3x - 4$$
$$\text{Expression B: } 12x - 4$$
$$\text{Expression C: } 5x + x - 4$$

Brianna says that all three expressions are equivalent because the value of each one is -4 when $x = 0$.

Brianna's thinking is incorrect.

Identify the error in Brianna's thinking. Determine which of the three expressions are equivalent. Explain or show your process in determining which expressions are equivalent.

Enter your answer and your explanation or process in the space provided.

▾ Math symbols

+	−	×	÷
±	‾	·	/
=	≠	▯/▯	▯▯
y^x	√	∛	π
(·)	°	|·|	

▸ Relations

▸ Geometry

Type 3: Modeling in Real-World Situations

This may involve diagrams, illustrations, or graphs. Questions may make use of place-value cubes. However, it might simply be designing an equation to fit a real-life scenario.

The graph shows the location of point *P* and point *R*. Point *R* is on the *y*-axis and has the same *y*-coordinate as point *P*.

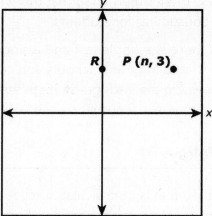

Point *Q* is graphed at $(n, ^-2)$. The distance from point *P* to point *Q* is equal to the distance from point *P* to point *R*.

What is the distance from point *P* to point *Q*? What is the value of *n*? Explain how you determined the distance from point *P* to point *Q*, and the value of *n*.

Enter your answers and your explanations in the space provided.

Type 1 questions are all computer scored, and Type 2 and Type 3 may be scored by a computer or a person. Different questions will be assigned a different weight (ranging from 1 to 6 points) depending on the number of parts needed to fully complete the task.

Changes Since the First PARCC Test

The first year of PARCC the students completed the assessment in two separate sections. The mid-year assessment, also known as the Performance-Based

Assessment (PBA), had three math sections (typically given over two days) in March and the End-of-Year Assessment (EOY) given in two sections (typically on the same day) in May. It has been determined that this was complicated for school districts to plan and time-consuming for students.

The current plan is that there is a single test and a single testing window. The window is from mid-April to mid-May, and schools and school districts will be able to plan their testing window. On the sixth-grade math test there are now three sections that are each 80 minutes in length for a total of 4 hours of math testing.

Calculator Usage Policy

On the test, there is a section that is noncalculator active and a section that allows you to use a calculator. The noncalculator section will be first. You must check your work in this section carefully before moving on to the calculator section. After you move forward to the next portion, you will NOT be able to go back to the noncalculator questions. They will not appear in a final review. Be sure to double check any work in the noncalculator section before moving on.

An icon will appear at the top as part of the tool bar next to the protractor on the section where you can use a calculator. It is a basic four function calculator. Your school may also choose to provide you with a separate four function calculator during this section of the test.

Keep in mind that, unlike the calculators you use in school, these do not follow order of operations or do any operations in fractions. They only perform one mathematical operation at a time. You will have to be sure that you follow mathematical rules.

Standards for Mathematical Practice

In addition to grade-specific content standards, the Common Core has also adopted Standards for Mathematical Practice that go across the grade levels. These eight standards have a context at each grade. The focus of these standards is the process, or the "how" of math instead of the product or answer. The idea is that if you interact with math in a variety of ways, beyond just memorizing rules, you allow for deeper and more long-lasting understanding of the concepts. These eight standards are

1. Make sense of problems and persevere in solving them

2. Reason abstractly and quantitatively

3. Construct viable arguments and critique the reasoning of others

4. Model with mathematics

5. Use appropriate tools strategically

6. Attend to precision

7. Look for and make use of structure

8. Look for and express regularity in repeated reasoning

These standards are addressed on the PARCC in the types of questions that will be asked. For example, Type 3 questions tie directly with standard 4. It is important that you are able to demonstrate your mathematical thinking in different ways in order to answer the questions posed to you.

Score Reports

PARCC score reports are designed to help both parents and teachers see where test takers show strengths and weaknesses so that gaps can be filled and enrichment provided to best educate students. PARCC has determined that this is best shown through an online report for teachers and an easy to understand printed report for parents.

The report sent home will show where the student is in the attainment of the grade level skills needed to be successful in school, and be on the right path to being ready for college and career at the time of graduation.

There are three major sections of the printed score report. The first section gives a score from 1–5, a 1 being minimal understanding and a 5 being distinguished understanding. The next section gives a three digit numerical score that can be used for comparison. Here a student's score will be compared to the average score received by other students in the school, in the district, in the state, and against all students taking the PARCC. Section three gives some details about how a student performed in comparison to peers that achieved the same 1-5 score. It will indicate if the student was below average, near average, or above average for his or her score group.

Fifth-Grade Review

This chapter will focus on some of the important concepts from fifth grade. These will be the skills you will need to master before moving on to the sixth-grade material. This chapter can either be completed entirely or be used as a reference if you are having difficulty with one of the new concepts.

Multiplication and Division

You have been learning your multiplication facts since at least third grade, and you will see that they will continue to be important as you move forward in your mathematics education. Multiplication is one of the most basic building skills, so it is important to emphasize, again, its significance. Additionally, you want to continue being confident in your division facts. Besides knowing how to multiply and divide whole numbers, you simply need to know the steps and work neatly so that your work stays lined up according to place value.

In multi-digit multiplication, we need to make sure that we multiply every digit in the first factor by every digit in the second factor.

For example,

$$\begin{array}{r} 1{,}234 \\ \times\ 65 \end{array} \quad \text{is really} \quad \begin{array}{r} 1{,}234 \\ \times\ 60 \end{array} \quad \text{added to} \quad \begin{array}{r} 1{,}234 \\ \times\ 5 \end{array}$$

The traditional way to multiply large digit numbers is to use layers. There should be one layer for each digit in the second factor.

$$\begin{array}{r} 1{,}234 \\ \times\ 65 \\ \hline 6{,}170 \\ +74{,}040 \\ \hline 80{,}210 \end{array}$$

6,170   This is the product of $1{,}234 \times 5$.

+74,040   This is the product of 1,234 and 60 or $1{,}234 \times 6$ with a zero at the end.

80,210   Our finished answer

When multiplying in traditional layers it is very important to keep your digits lined up as you go. If you have an answer of 0 (zero), or with a zero in the ones place, be sure that you put the zero on the answer line.

For every layer added after the first, there is an additional zero placed on the right side of the answer. These "place holder zeros" move your answer over to the left. The second line has one place holder zero because you are multiplying by a factor of 10. The third line has two zeros because you are multiplying by a factor of 100. The fourth line would have three zeros and so forth.

You will also find that as you progress in this book there will start to be other ways to write multiplication. Up until now multiplication was signified by a "×" and the factors were either next to each other or written vertically. There are other symbols for multiplication including a dot and parentheses.

For instance, 3 × 4 is the same as 3 · 4 is the same as 3(4). When we start adding variables, putting them right next to each other also means multiply. $3y = 3 \times y$.

Likewise, division has multiple symbols. Division can be marked with "÷" or in long form using $\overline{)}$ or as a fraction.

In dealing with long division, we have several steps that we have to remember.

You may have used the following to help you remember the steps:

Does **M**cDonalds **S**ell **B**urgers   or   **D**aughter **M**other **S**ister **B**rother

These mnemonic devices stand for **D**ivision, **M**ultiplication, **S**ubtraction, and **B**ring down. Each letter is the step in order that we follow, repeating for as many digits are in the dividend.

$$
\begin{array}{r}
173 \\
8\overline{)1{,}386} \\
-8 \\
\hline
58 \\
-56 \\
\hline
26 \\
-24 \\
\hline
2
\end{array}
$$

As we move through the digits, we address each place value. In this example, we first look at groups of 1,000, and we find that we don't have enough. We move to make groups of 100; we had 1 group and 5 left. These left over get broken into groups of 10 and so forth. The part left over when there is nothing left to bring down is called the remainder.

The **remainder** can be written as a whole remainder or as a fraction. The remainder is the numerator and the original divisor is the denominator. The initial divisor is the numerator because that was the number of sections we originally created.

$$\frac{173 \text{ R2}}{8 \overline{)1{,}386}}$$

$$\text{Whole} \quad \frac{\text{Numerator (Remainder)}}{\text{Denominator (Divisor)}} = 173\frac{2}{8} = 173\frac{1}{4}$$

When dividing with zeros you have to follow special rules.

- If zero is your dividend your answer is always zero. No matter how many groups you have, if you start with zero, each group will have zero in it.
- If zero is your divisor...STOP! You cannot divide by zero! This will get you an error on your calculator.

Practice

1. 5,472 × 655
2. 1,236 × 2,178
3. 52,136 ÷ 29
4. 61,224 ÷ 43

Notebook

Answers

1. **3,584,160**
2. **2,692,008**
3. **1,797 R 23**
4. **1,423 R 35**

Properties and Order of Operations

There are four main properties that you should have already dealt with in fifth grade. The first three properties are **Identity**, **Commutative**, and **Associative**. They all work with both addition and multiplication. The **Zero Property** is just for multiplication.

The **Commutative Property** states that it doesn't matter what order the numbers are in as long as we are either multiplying or adding all of them.

3 + 2 + 5 is the same as 5 + 2 + 3 or 2 + 3 + 5. We always get 10.

The same is true of multiplication.

3 × 6 × 2 = 36 and 2 × 3 × 6 = 36 and 6 × 2 × 3 = 36.

All arrangements of these numbers will give the same value. This is particularly important when trying to find the simplest way to solve a problem. For example, let's look at the problem $6 \times 3 \times 3$. If we solve 6×3 first we get 18. However, since we may not know as mental math what 18×3 equals, if we rearrange the order $3 \times 3 \times 6$, we know that 3×3 is 9 and 9×6 is 54. This means that 18×3 would also equal 54. There... we found an easier way!

 HELPFUL HINT

The Commutative Property does NOT work for subtraction, division, or when addition and subtraction are mixed.

The **Identity Property** can also be applied to both addition and subtraction. The Identity Property of addition or subtraction states that adding or subtracting zero and any number results in an answer identical to the original number.

$$6+0=6 \qquad \frac{3}{7}+0=\frac{3}{7} \qquad 1,713,124+0=1,713,124 \qquad 5,621-0=5,621$$

When we add or subtract zero it keeps the original number, or identity. Multiplication and division works the same way except instead of 0, we multiply or divide by 1.

$$6\times1=6 \qquad \frac{3}{7}\times1=\frac{3}{7} \qquad 1,713,124\times1=1,713,124 \qquad 5,215\div1=5,215$$

The **Associative Property** is working with parentheses. Like the Commutative Property, it is saying that the order of addition or multiplication does not matter. If there is a series of addition or a series of multiplication, we can use parentheses to group the numbers differently and get the same answer.

$2 \times (5 \times 6)$ is the same as $(2 \times 5) \times 6$; both will give an answer of 60. $(2 + 6) + 8$ is equivalent to $2 + (6 + 8)$ because both will result in a total of 16.

The **Zero Property** is sometimes mistaken for the identity property. The Zero Property is the result of multiplying any factor times zero. It ALWAYS equals zero.

$$6\times0=0 \qquad \frac{3}{7}\times0=0 \qquad 1,713,124\times0=0$$

This makes sense because no matter how many boxes (groups we have), if they are all empty (0), we have nothing. Remember that because of the Commutative Property, zero times any number also equals zero.

Order of Operations

You have already practiced with order of operations: **P**lease **E**xcuse **M**y **D**ear **A**unt **S**ally

Parentheses

Exponents

Multiplication Division

Addition Subtraction

Notice that we kept multiplication and division on one line and addition and subtraction together on another. There is an important reason! Multiplication and division are done at the same time (from left to right). After, we complete the addition and subtraction together, from left to right.

Order of operations is extremely important in order to get correct and accurate answers.

Let's look at an example.

Bobby and Rose are completing a multi-step math problem:

$$2 + 4 \times 6$$

Bobby works from left to right:

$$2 + 4 = 6 \times 6 = 36$$

Rose followed order of operations:

$$4 \times 6 = 24 \quad 2 + 24 = 26$$

Both had correct computations but got very different answers. Rose's answer is correct because she used the correct order of operations.

Peter and David both have a good understanding of order of operations and know that they need to multiply and divide before adding. They still got different answers on the problem $3 + 8 \div 2 \times 2$.

Peter solved the problem:	David solved the problem:
$3 + 8 \div 2 \times 2$	$3 + 8 \div 2 \times 2$
$3 + 8 \div 4$	$3 + 4 \times 2$
$3 + 2 = 5$	$3 + 8 = 11$

David has the answer correct. We do not necessarily solve multiplication before we solve division. We solve them in the order they appear from left to right, which is how David approached the problem.

Practice

Identify the property.

1. $4 \times 0 = 0$ *Zero Property*
2. $4 \times 3 \times 2 = 2 \times 3 \times 4$ *Comumitve*
3. $5 + 0 = 5$ *Identity*
4. $5 + (2 + 3) = (5 + 2) + 3$ *Associtive*

Solve the following order of operations.

5. $3 + 6 \times (5 - 3)$ $3 + 6 \times (5-3)$ $7^2 - 9 \div 3 \times 2$

   $3 + 6 \times (3)$ $49 - 9 \div 3 \times 2$

6. $7^2 - 9 \div 3 \times 2$ $3 + 18$ $49 - 3 \times 2$

   21 $49 - 6$

   43

Answers

1. **Zero Property**

2. **Commutative Property**

3. **Identity Property**

4. **Associative Property**

5. $3 + 6 \times (5 - 2)$ 6. $7^2 - 9 \div 3 \times 2$

   $3 + 6 \times (3)$ $49 - 9 \div 3 \times 2$

   $3 + 18$ $49 - 3 \times 2$

   21 $49 - 6$

                                   43

Prime and Composite Numbers

Many times when working with numbers we are looking at various ways that we can break them apart or combine them. Often this is achieved using **prime numbers**. A prime number is a number that only has two factors, 1 and itself. In other words, this is a number that is not evenly divisible. **Composite numbers** are made of several different pairs of factors. These factor pairs still include 1 and itself, but there are other ways to break the number down into groups of whole numbers.

For example:

7 is a prime number. We cannot divide 7 by any other whole number and get a whole number answer. The only factors of 7 are 1 and 7.

12 is a composite number. It has the factors 1 and 12, but can also be made of 2 and 6, or 3 and 4.

By definition, the numbers 1 and zero are neither prime nor composite. The number 1 has only one factor, not two. Zero has infinite factors because zero times anything equals zero.

1	2	3	4	5	6	7	8	9	10
11	12	13	14	15	16	17	18	19	20
21	22	23	24	25	26	27	28	29	30
31	32	33	34	35	36	37	38	39	40
41	42	43	44	45	46	47	48	49	50
51	52	53	54	55	56	57	58	59	60
61	62	63	64	65	66	67	68	69	70
71	72	73	74	75	76	77	78	79	80
81	82	83	84	85	86	87	88	89	90
91	92	93	94	95	96	97	98	99	100

If you take the above 100s chart and cross out all your multiples, whatever numbers remain are your prime numbers.

Directions: Circle the number 2. Now cross out all other multiples of two (4, 6, 8, 10, 12....). Repeat with circling 3. Now cross out all other multiples of 3 (6, 9, 12, 15, 18....). Do this for all of your factors.

After you finish you will be left with all your prime numbers circled. Your prime numbers up to 100 are 2, 3, 5, 7, 11, 13, 17, 19, 23, 29, 31, 37, 41, 43, 47, 53, 59, 61, 67, 71, 73, 79, 83, 89, and 97.

Now that we have identified our prime numbers we can start looking at ways to break apart composite numbers. The first is to identify factor pairs. Factor pairs are two numbers that are multiplied to make a given value. Composite numbers need to have at least one factor pair other than one and itself, but they can have many factor pairs.

When listing factor pairs it is important to go in order, as to not miss any pairs. Once you meet in the middle you have found all the pairs.

1,	, 40	1 × 40
1, 2,	20, 40	2 × 20
1, 2, 4,	10, 20, 40	4 × 10
1, 2, 4, 5, 8, 10, 20, 40		5 × 8

Sometimes we want to break a composite number into its smallest possible pieces. Our smallest pieces are prime numbers. Some composite numbers are made of one factor pair of two prime numbers. For example 6 is 3 × 2, both 3 and 2 are prime. Other numbers are more complex. For example 18 can be thought of as 9 × 2. The number 2 is a prime number but 9 is not. We need to break 9 down further. We know that 9 is 3 × 3. This means that if we multiply the prime numbers 3 × 3 × 2 we get a total of 18. The easiest way to do this is a **factor tree,** or factor firework.

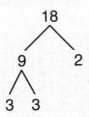

The number 18 is made of 3 and 3 and 2: 2 × 3 × 3 = 18. The list of the prime numbers is called the **prime factorization** or **product of primes**. When finding the primes it doesn't matter what factor pair you start with, you will end up with the same prime numbers.

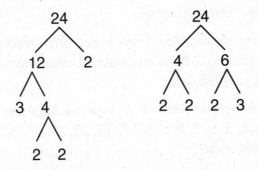

No matter which way we start 24 we end up with 2 × 2 × 2 × 3 = 24.

When listing the prime factorization we always list the primes in order from least to greatest. For example, we do not write 2 × 3 × 2 × 2, but instead 2 × 2 × 2 × 3. You can also write the prime factorization using exponents. The prime factorization for this problem can be written as $2^3 × 3$. You will learn more about exponents in chapter 5.

Practice

Determine if the following are prime or composite. If composite list the prime factorization.

1. 17 *Prime*
2. 24 *composite 2 x 2 x 2 x 3*
3. 145 *composite 5 x 29*
4. 87 *composite 3 x 29*
5. 71 *Prime*

Answers

1. Prime
2. Composite $2 \times 2 \times 2 \times 3$
3. Composite 5×29
4. Composite 3×29
5. Prime

Equivalent Fractions

Equivalent fractions are fractions that are comprised of different digits but have the same value. For example $\dfrac{1}{2}$ and $\dfrac{2}{4}$ are equivalent; they represent the same amount.

$$\frac{1}{2} \quad = \quad \frac{2}{4} \quad = \quad \frac{3}{6} \quad = \quad \frac{4}{8}$$

Multiplying any number times 1 doesn't change the value.

$$6 \times 1 = 6$$

$$47 \times 1 = 47$$

This is also true of fractions.

$$\frac{1}{6} \times 1 = \frac{1}{6}$$

We also know that in fractions when the numerator and denominator are the same number the value of the fraction is one whole. If I have 8 pieces and use 8 pieces I used the whole.

$$\frac{8}{8} = 1$$

Think about this as a magic one.

A magic one can have any fraction with the same numerator or denominator inside. We are really just multiplying by 1.

$$\frac{5}{2} \times \boxed{1 \frac{4}{4}}$$

To make equivalent fractions, we simply multiply or divide both the numerator and the denominator by the same value.

$$\frac{1}{2} \frac{\times 2}{\times 2} = \frac{2}{4} \frac{\times 3}{\times 3} = \frac{6}{12}$$

When we take a fraction and divide both the numerator and the denominator by the greatest common factor, we end up with a special equivalent fraction called **simplest form.** We get simplest form when we divide the digits until we can't make the digits any smaller and keep them whole numbers. When working with fractions, answers should be reduced to the simplest form.

$$\frac{12}{16} \frac{\div 4}{\div 4} = \frac{3}{4}$$

Practice

Which fraction in the line is not equivalent?

1. $\dfrac{2}{3}$ $\quad\dfrac{10}{15}$ $\quad\boxed{\dfrac{8}{9}}$ $\quad\dfrac{16}{24}$ $\quad\dfrac{4}{6}$

2. $\boxed{\dfrac{5}{10}}$ $\quad\dfrac{7}{49}$ $\quad\dfrac{2}{14}$ $\quad\dfrac{5}{35}$ $\quad\dfrac{4}{28}$

Answers

1. $\dfrac{8}{9}$

2. $\dfrac{5}{10}$

Comparing and Ordering Fractions

You may, at times, need to be able to put fractions in order from least to greatest, or use comparison symbols such as < or >. The mathematical way to approach this problem is to make all the denominators the same. When your denominators are all the same it is easy to put the numerators, and therefore the fractions, in order. By using the rules of equivalent fractions we can find equivalent fractions that result in common denominators.

For example: Which is larger, $\dfrac{2}{5}$ or $\dfrac{1}{4}$? Our first decision is what denominator can we make from both fifths and fourths? The answer is 20. So we need to make the equivalent fractions.

$$\frac{2}{5} = \frac{?}{20}$$

How do we get from 5 to 20? Multiply by 4! we need to multiply the numerator by 4 as well. $2 \times 4 = 8$.

The equivalent fraction is $\dfrac{8}{20}$.

Now repeat the same thing for $\dfrac{1}{4}$.

$$\frac{1}{4} = \frac{?}{20}$$

How do we get from 4 to 20? Multiply by 5! So we need to multiply the numerator by 5 as well. $1 \times 5 = 5$.

The equivalent fraction is $\dfrac{5}{20}$.

We can now compare fractions with the same denominator.

$\dfrac{8}{20}$ is larger than $\dfrac{5}{20}$.

$$\frac{8}{20} > \frac{5}{20}, \text{ so } \frac{2}{5} > \frac{1}{4}$$

The Zig-Zag Method

When you have two fractions there is an easier way to compare them. This method has lots of silly names. You may have heard of the butterfly method or the zig-zag method. Basically we are finding diagonal products.

We need to multiply the numerator of each fraction by the other fraction's denominator. Which is exactly what we did in the previous section.

Which is larger, $\frac{3}{11}$ or $\frac{4}{15}$? We can use zig-zag to determine. The important part of the zig-zag method is that we keep the product with the numerator.

$$\frac{3}{11} \quad \frac{4}{15}$$
$$3 \times 15 = 45 \qquad 4 \times 11 = 44$$

Since 45 is larger than 44, that means that $\frac{3}{11}$ is larger than $\frac{4}{15}$. We really found the numerators for the equivalent fractions. We don't need to find the corresponding denominator (which would be $11 \times 15 = 165$) because the numerators determine which is larger.

 HELPFUL HINT

If you have two fractions with the same numerator but different denominators you can also compare them. The larger denominator is the smaller fraction, for instance $\frac{2}{5}$ is smaller than $\frac{2}{3}$.

Estimation and Number Lines

There are times that we do not need to find equivalent fractions. When you are given a set of fractions that need to be put in order from least to greatest, you may be able to put many of these fractions in order without recalculating them. Try to separate numbers that are closest to zero (think a 1 in the numerator), around $\frac{1}{2}$, and closest to 1 whole (the numerator is close in value to the denominator). This may allow you to separate your values into smaller, more manageable groups, or even determine their order. Drawing a number line can help organize your fractions.

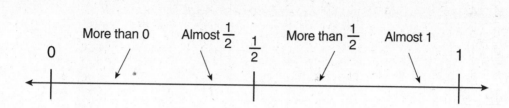

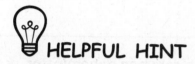 **HELPFUL HINT**

It can be beneficial to know your equivalent fractions to one half. Your denominator is always twice the numerator ($\frac{1}{2}$, $\frac{2}{4}$, $\frac{3}{6}$, $\frac{7}{14}$ ...).

Practice

Compare using <, > sign.

1. $\frac{9}{12} > \frac{3}{5}$

2. $\frac{4}{8} < \frac{4}{5}$

3. $\frac{2}{5} > \frac{1}{3}$

4. $\frac{15}{9} < \frac{21}{2}$

Put the following fractions in order from least to greatest.

$$\frac{8}{9}, \ 1\frac{1}{6}, \ 1\frac{7}{9}, \ \frac{1}{6}, \ \frac{13}{19}$$

$\frac{1}{6}, \frac{13}{19}, \frac{8}{9}, 1\frac{1}{6}, 1\frac{7}{9}$

Answers

1. $\dfrac{9}{12} > \dfrac{3}{5}$

2. $\dfrac{4}{8} < \dfrac{4}{5}$

3. $\dfrac{2}{5} > \dfrac{1}{3}$

4. $\dfrac{15}{9} < \dfrac{21}{2}$

$$\dfrac{1}{6}, \dfrac{13}{19}, \dfrac{8}{9}, 1\dfrac{1}{6}, 1\dfrac{7}{9}$$

Adding and Subtracting Fractions

It is easy to add and subtract fractions that have a **common denominator,** or the same bottom number.

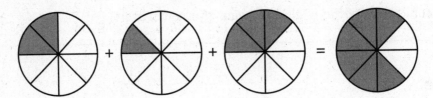

The denominator is like a label.

1 kitten plus 1 kitten is 2 kittens

1 eighth plus 1 eighth is 2 eighths

$$\dfrac{1}{8} + \dfrac{1}{8} = \dfrac{2}{8}$$

If we end up with an **improper fraction,** or a numerator that is larger than a denominator, we need to make a **mixed number**. We change an improper fraction to a mixed number by finding how many wholes we have in the fraction and what fraction is remaining. Remember that when the numerator and denominator are the same ($\dfrac{2}{2}$, $\dfrac{3}{3}$, $\dfrac{4}{4}$), we have one whole.

$$\dfrac{8}{3} = \dfrac{3}{3} + \dfrac{3}{3} + \dfrac{2}{3} = 2\dfrac{2}{3}$$

Subtraction is easy too. It works just like addition, except we subtract the numerator.

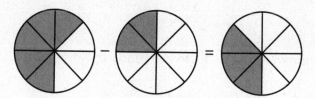

Subtraction can be a challenge when dealing with mixed numbers. Sometimes there will not be enough in the fraction portion to subtract. Just as in whole number subtraction, we will have to borrow. When we borrow, we take one whole and split it into fractions. Create the fraction based on the denominator you are trying to subtract.

$$3\frac{1}{6} = 2 + \frac{6}{6} + \frac{1}{6} = 2\frac{7}{6}$$

After we borrow, we now will have enough to subtract.

$$3\frac{1}{6} - 2\frac{4}{6} = 2\frac{7}{6} - 2\frac{4}{6} = \frac{3}{6} = \frac{1}{2}$$

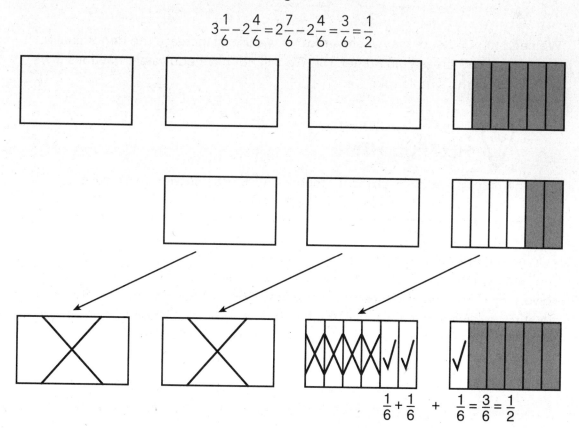

$$\frac{1}{6} + \frac{1}{6} \quad + \quad \frac{1}{6} = \frac{3}{6} = \frac{1}{2}$$

Unfortunately, we don't always deal with fractions that have the same denominator. Sometimes they are different, and we first have to use equivalent fractions to find a common denominator before we can proceed to addition and subtraction.

$$\frac{1}{4} + \frac{1}{2}$$

First we have to find a common denominator. In this case, we are lucky and only have to change the half.

$$\frac{1}{2} = \frac{2}{4}$$

Now that both denominators are 4, we are ready to add.

$$\frac{1}{4} + \frac{2}{4} = \frac{3}{4}$$

Sometimes we are not so lucky and can't change one denominator to match the other. We have to find equivalent fractions for both.

$$\frac{2}{3} - \frac{1}{4}$$

Here we have to look for a number that is a multiple of both 3 and 4. The first one we find is 12.

$$\frac{2\times4}{3\times4} = \frac{8}{12} \qquad \frac{1\times3}{4\times3} = \frac{3}{12} \qquad \frac{8}{12} - \frac{3}{12} = \frac{5}{12}$$

We need to make sure that whatever factor we use to increase the denominator, we also use to increase the numerator. We will not have equivalent fractions if we do not follow this simple rule.

 HELPFUL HINT

After you add or subtract, double-check your answer to be sure that you cannot simplify the fraction.

Finding the **lowest common denominator** means finding the smallest multiple of both denominators. We don't have to use the smallest common denominator to solve. Any common denominator will work: we will just have more simplifying to do at the end. If we multiply each fraction by the opposite denominator, we will have a common denominator. This is called **cross-multiplication.**

$$\frac{3}{4} - \frac{4}{6} \qquad \frac{3\times6}{4\times6} = \frac{18}{24} \qquad \frac{4\times4}{6\times4} = \frac{16}{24} \qquad \frac{18}{24} - \frac{16}{24} = \frac{2}{24} = \frac{1}{12}$$

The lowest common denominator is 12. We get the same answer by cross-multiplying and simplifying.

Practice

1. $1\frac{2}{9} + 2\frac{5}{12} =$ $3\frac{23}{36}$

2. $3\frac{1}{5} - 1\frac{7}{10} =$ $1\frac{5}{10} = 1\frac{1}{2}$

Answers

1. $3\frac{23}{36}$

2. $1\frac{5}{10} = 1\frac{1}{2}$

Multiplying Fractions

When you first learned to multiply, you probably thought it was harder than adding. However, you will find that is probably not true of fractions! It can be tricky to find common denominators in order to add and subtract fractions. When you multiply, you get to skip that step!

STEP 1 Multiply the numerators of both fractions.
STEP 2 Multiply the denominators of both fractions.
STEP 3 Be sure fraction is in simplest form.

Easy!

$$\frac{2}{3} \times \frac{5}{6} = \frac{2\times5}{3\times6} = \frac{10}{18} \quad \text{simplified to } \frac{5}{9}$$

If you are working with mixed numbers, you have to make sure that you first change them into improper fractions.

$$3\frac{2}{5} \times 1\frac{4}{6} = \frac{17}{5} \times \frac{10}{6} = \frac{170}{30} = 5\frac{20}{30} = 5\frac{2}{3}$$

Be sure that you keep all your work neat!

Practice

1. $4\frac{2}{5} \times 2\frac{1}{9}$

2. $\frac{4}{7} \times \frac{5}{4}$

3. $3\frac{9}{10} \times \frac{2}{3}$

Notebook

Answers

1. $9\frac{13}{45}$

2. $\frac{20}{28} = \frac{5}{7}$

3. $\frac{78}{30} = 2\frac{18}{30} = 2\frac{3}{5}$

Decimal Place Value

As you have gotten older you have learned the importance of place value. The value of a 6 is much larger when it is put in the tens place (60) or the hundreds place (600) and the same is true of working on the other side of the decimals.

10 tenths = 1 whole

10 hundredths = 1 tenth

Thousands	Hundreds	Tens	Ones	Decimal Point	Tenths	Hundredths	Thousandths
				●			

HELPFUL HINT

When reading a decimal we say the whole number (just as we normally would), then we say "and" followed by the decimal. The decimal is named for its place value.

Decimals, like fractions, are representatives of a part of a whole. The fraction $\frac{1}{10}$ is equal to the decimal 0.1. Both of these representations show that this is one part, out of 10 equal parts, that make up one whole. Decimals are most frequently seen in money. Money is always to two decimal places or the hundredths place. The decimal portion of the money is the part less than a whole dollar, the coins. For instance, $10.37 is 10 whole dollars and .37 of a dollar. We can say it is thirty-seven hundredths of a dollar. Remember there are 100 cents in one whole dollar.

Decimals are also seen in measurements. For example, the distance from your house to your school is 3.7 miles. That means you would need to walk more than 3 miles but not quite 4 miles to make the trip. The .7 tells us that if I split the mile into ten equal pieces you have to walk 7 of them.

Remember:

Multiplying by 10 moves your decimal over one place to the right.

Multiplying by 100 moves your decimal over two places to the right.

Multiplying by 1,000 moves your decimal over three places to the right.

Dividing by 10 moves your decimal over one place to the left.

Dividing by 100 moves your decimal over two places to the left.

Dividing by 1,000 moves your decimal over three places to the left.

Comparing and Ordering Decimals

Just like in whole numbers, the place value in a decimal is very important. Just like you know that 4 is smaller than 40 we need to know the difference between 0.4 and 0.04. Both 0.4 and 0.04 represent 4 parts; the difference is the size of the part. 0.4 is 4 tenths; we took one whole and split it into 10 and are using 4 of them. 0.04 is 4 hundredths; we took one whole and split it into 100 and are using 4 of them. The more pieces we cut from the same whole, the smaller their pieces. This means that 0.04 is 4 much smaller pieces than 0.4.

$$0.4 > 0.04$$

HELPFUL HINT

Moving a digit to the left increases its value, moving it to the right decreases its value.

Practice

Compare the decimals using <, >, or =.

1. 3.2 > 3.02

2. 4.1 < 41

3. 3.08 = 3.080

4. 6.08 < 6.2

5. 0.07 < 0.7

Answers

1. 3.2 > 3.02

2. 4.1 < 41

3. 3.08 = 3.080

4. 6.08 < 6.2

5. 0.07 < 0.7

Perimeter and Area of Rectangles

Sixth-grade geometry is mostly measuring figures. We should start by reviewing **perimeter**. Perimeter is the measurement around the outside of a figure. This is often discussed as fencing on a field, border in a room, or a frame around a picture. It is found by adding all sides together. In a triangle with 3 sides we add side + side + side. A rectangle has 4 sides so we add side + side + side + side.

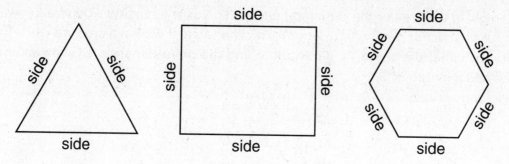

In a rectangle opposite sides are equal so instead of adding

side + side + side + side

We call one direction length and the other direction width

length + length + width + width   or   (length + width) × 2

Don't forget to label your answers. If no label is provided, use the label **units**.

Area is measuring how many **square units** can fit in a given space. Area of a rectangle is just like when you learned to multiply using an array. It is length times width. In other words, area is calculating the number of rows and columns of squares.

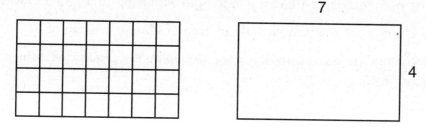

Finding area is often a discussion about how much carpet is needed in a room, tile on a floor, size of a garden, or size of a classroom bulletin board. If you can find area with the formula length × width = area, you can find the length using area ÷ width, and find width using area ÷ length. Imagine if you had a length of 7 units and a width of 3 units. This would give you an area of 21 square units. If you have an area of 21 square units and a length of 7 units, that leaves you with a width of 3 units. The numbers 3, 7, and 21 create a fact family.

When given the area or the perimeter we can solve for a missing side. We know that for a rectangle we multiply *length* × *width* = *area*. If we have area we can divide by either the length or the width to find the other dimension. Given an area of 27 and a length of 3,

$$3 \times w = 27$$

$$27 \div 3 = 9$$

The missing width is 9.

We can use subtraction for perimeter. If we know that two sides of a triangle are 3 and 6 and the total perimeter is 14, we can subtract to find the missing side.

$$14 - 3 = 11$$

$$11 - 6 = 5$$

The missing side has a length of 5.

Practice

1. Find the perimeter and area for a rectangle with a length of 7 and width of 3.
2. Find the perimeter and area for a rectangle with a length of 35 and width of 17.
3. Find the length of a rectangle with an area of 45 and a width of 3.
4. Find the length of each side of a pentagon with all equal sides if the total perimeter is 40.

Answers

1. **Perimeter = 7 + 3 + 3 + 7 = 20 units**

   Area = 7 × 3 = 21 square units
2. **Perimeter = 35 + 35 + 17 + 17 = 104 units**

   Area = 17 × 35 = 595 square units
3. **3 × ? = 45**

   45 ÷ 3 = 15

A pentagon has 5 sides. If they are all equal we can divide by 5.

$$40 \div 5 = 8$$

Graphing in a Coordinate Plane

If you have ever played the game of *Battleship* or completed a hidden picture, then you have worked with a coordinate plane. A coordinate plane is an organized way of plotting points. Each point has an "address" determined by how far across (the *x* value) and how far up (the *y* value) should be positioned on the chart.

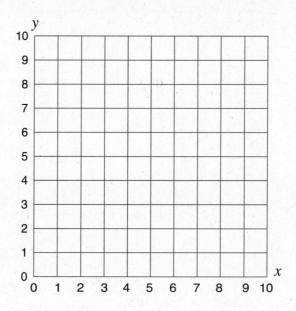

This "address" is called a coordinate point and it is a single spot on the coordinate plane. Coordinate points are written in pairs inside parentheses with a comma separating the *x* value and the *y* value.

$$(x, y)$$

Notice that the *x* value comes first. (The across *always* comes first.) It is easy to remember A comes before U in the alphabet so we go ACROSS before we go UP.

Between the lines there are other addresses; for example $(1\frac{1}{2}, 3\frac{1}{2})$ is an address on this plane, and it is just between lines.

Practice

Plot the points (3, 2), (4, 5), and (7, 1).

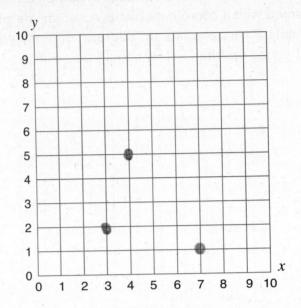

Answers

Notice it is always the *x* number before the *y*.

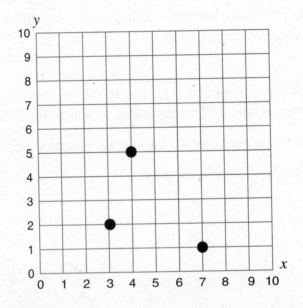

Review Test

1. Delilah and her soccer club are raising money to replant the grass on their field. They raised a total of $22,446. There are 58 people in the club. What is the average amount that each person raised? =387 B

2. There are 68 books that the librarian wants to divide equally on the shelves. List all the possible ways she can arrange the books in equal groups.

3. Solve the following: *PEMDAS*

$$12 + 5 \times 5^2 - (6 \div 2)$$

4. Jude was working on his homework. He had to solve the problem $3^2 + 6 \div 3$ and got an answer of 5. What was his mistake?

5. What is the prime factorization of 235? = 5×47

6. Put the fractions in order from least to greatest.

$$\frac{2}{3}, \ \frac{1}{7}, \ \frac{2}{5}, \ \frac{3}{7}, \ \frac{1}{5}$$

7. What is $3,456 \div 24$? = 144

8. Francesca is making necklaces to sell at craft fairs. Each necklace needs 315 beads. She wants to make 125 for the summer. How many beads does she need? 39,375

9. $1\frac{3}{4} + 2\frac{3}{5} =$ 4 $\frac{7}{20}$

10. $5\frac{1}{6} - 3\frac{3}{8} =$ 1 $\frac{19}{24}$.

11. $1\frac{2}{3} + 2\frac{1}{5} + 3\frac{3}{4} =$ 7 $\frac{37}{60}$

12. Find the perimeter and area of a square with sides that are $2\frac{1}{2}$ inches. =10

13. Elena has two different rectangles that have the same perimeter of 24 inches. Their sides are measured in whole inches. What are two possible rectangles that she has? Do they have the same area? Explain.

14. Tyler had $\frac{3}{4}$ of a pizza. The pizza originally had 12 slices. How many slices did Tyler eat? = 9

15. Plot the points on the coordinate plane: (3, 6) (1$\frac{1}{2}$, 5) and (6, 2$\frac{1}{2}$).

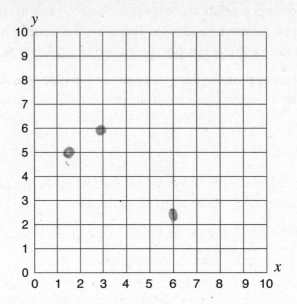

16. Which of the following fractions are greater than $\frac{1}{3}$?

   ● A. $\frac{2}{5}$

   ● B. $\frac{5}{10}$

   ○ C. $\frac{5}{15}$

   ○ D. $\frac{6}{20}$

   ● E. $\frac{9}{3}$

17. Barry makes 8 dollars an hour. He has been working for 5$\frac{3}{4}$ hours. How much did Barry earn? 46$

18. What is the value of the 7 in each of the following numbers?

   27.3

   13.7

   214.307

   10.7

   13.27

Answers

1. Simple Division—Each student raised an average of $387.

2. Finding Factor Pairs—One shelf of 68 books; 68 shelves of 1 book; 2 shelves of 34; 34 shelves of 2; 4 shelves of 17; and 17 shelves of 4.

3. $12 + 5 \times 5^2 - 3$

    $12 + 5 \times 25 - 3$

    $12 + 125 - 3$

    $137 - 3$

    134

4. Improper order of operations. He worked left to right. $3^2 + 6 \div 3$ became $9 + 6 \div 3$, which he solved as $15 \div 3$ and an answer of 5. He was correct up to $9 + 6 \div 3$. At this point he should have divided first to get $9 + 2$. The correct answer is 11.

5. 5×47

6. $\dfrac{1}{7}, \dfrac{1}{5}, \dfrac{2}{5}, \dfrac{3}{7}, \dfrac{2}{3}$

7. 144

8. 39,375 beads

9. $4\dfrac{7}{20}$

10. $1\dfrac{19}{24}$

11. $7\dfrac{37}{60}$

12. Perimeter $2\dfrac{1}{2} \times 4 = 10$ inches

    Area $2\dfrac{1}{2} \times 2\dfrac{1}{2} = 6\dfrac{1}{4}$ square inches

13. You should have two of the possible answers, 1×11, 2×10, 3×9, 4×8, 5×7, or a 6×6. This creates areas of 11, 20, 27, 32, 35, and 36, respectively. You should realize that rectangles with the same perimeter may not have the same area.

14. We should know that we make this into equivalent fractions and the answer is 9.

15.

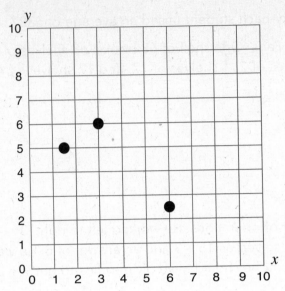

16. Only $\dfrac{2}{5}, \dfrac{5}{10}, \dfrac{9}{3}$

$\dfrac{5}{15}$ is equal not greater

17. $8 \times 5\dfrac{3}{4} = 46$ dollars.

18. 7 or 7 whole

7 tenths

7 thousandths

7 tenths

7 hundredths

The Number System

CHAPTER 3

The number system is looking at how numbers work. In previous years you have already mastered adding, subtracting, multiplying, and dividing whole numbers. Fractions have also been a big part of your math education. For the 6th Grade PARCC test we will be focusing on dividing fractions, calculating with decimals, and introducing negative numbers. The number system is the foundation for math understanding.

Division of Fractions

In the review section (Chapter 2), you learned a lot about fractions. At this point you should be comfortable adding, subtracting, and multiplying fractions.

Division of fractions itself is very simple. In fact, what we actually do is multiply! Multiplication is the reverse of division so if we multiply by the reverse, we will get the correct answer. In other words, we are multiplying by the reciprocal. What does that mean, you ask? We do the opposite of division (multiplication) with the opposite number (the reciprocal).

What is a **reciprocal**? It is an inverted or flipped fraction.

$$\frac{1}{2} \rightarrow \frac{2}{1}$$

$$\frac{3}{7} \rightarrow \frac{7}{3}$$

$$5 \rightarrow \frac{1}{5} \quad \text{Remember the denominator for any whole number is 1.}$$

$$1\frac{1}{3} = \frac{4}{3} \rightarrow \frac{3}{4} \quad \text{If we have a mixed number, we make an improper fraction and flip.}$$

Practice Finding Reciprocals

1. $\frac{4}{5} = \frac{5}{4}$

2. $\frac{2}{3} = \frac{3}{2}$

3. $6 = \frac{1}{6}$

4. $\frac{1}{8} = 8$

Answers

1. $\dfrac{5}{4}$

2. $\dfrac{3}{2}$

3. $\dfrac{1}{6}$

4. **8**

Once we have a reciprocal, we simply need to multiply! We multiply the numerator times the numerator and then the denominator times the denominator. Remember to put your answer in the simplest form.

$$\frac{2}{3} \div \frac{5}{7} \qquad \frac{2}{3} \times \frac{7}{5} = \frac{14}{15}$$

 HELPFUL HINT

It matters which fraction becomes a reciprocal; it is always the second one. Just remember Keep, Change, Flip!

Practice

1. $\dfrac{1}{3} \div \dfrac{2}{7} = \dfrac{1}{3} \times \dfrac{7}{2} = \dfrac{7}{6} = 1\dfrac{1}{6}$

2. $\dfrac{3}{4} \div \dfrac{3}{4} = \dfrac{3}{4} \times \dfrac{4}{3} = \dfrac{12}{12} = 1$

3. $5 \div \dfrac{3}{5} = \dfrac{5}{1} \times \dfrac{5}{3} = \dfrac{25}{3} = 8\dfrac{1}{3}$

4. $7 \div \dfrac{6}{11} = \dfrac{7}{1} \times \dfrac{11}{6} = \dfrac{77}{6} = 12\dfrac{5}{6}$

Answers

1. $1\dfrac{1}{6}$

2. 1

3. $8\dfrac{1}{3}$

4. $12\dfrac{5}{6}$

PARCC Question

The area of a rectangular field is $\frac{3}{5}$ square mile. The width of the field is $\frac{7}{8}$ mile.

What is the length of the field? Show your work.

Answer Explained

The area of a rectangle is L × W = A so we use the equation A ÷ L = W:

$$\frac{3}{5} \div \frac{7}{8} = \frac{24}{35} = \textit{keep Change flip}$$

Multi-Digit Division

Long division is a multi-step process that requires you to follow set steps and repeat as needed. You should remember this as Divide, Multiply, Subtract, Bring down (Repeat!).

You may have learned a fun way to remember these steps.

Does **M**cDonalds **S**ell **B**urgers

Daughter **M**other **S**ister **B**rother

Or you might put the symbols ÷ × − ↓ along the edge of the page.

Can you think of your own way to remember the steps? Write it here.

Divison, Multiplication, Subtraction, Bring down

Previously you have only had to divide with a single digit divisor.

$$
\begin{array}{r}
175 \\
5{\overline{\smash{\big)}\,875}} \\
-5\downarrow \\
\hline
37 \\
-35\downarrow \\
\hline
25 \\
-25 \\
\hline
0
\end{array}
$$

Now we need to figure out what to do when the divisor gets bigger.

$$2{,}688 \div 28$$

We start by setting up our long division problem.

$$28 \overline{)2{,}688}$$

Can we get a group of 28 out of 2? No.

What about a group of 28 out of 26? No.

Can we get groups of 28 from 268? Yes.

But how many?

It can be helpful to write a multiplication table for the divisor before you start. Remember you are working on one digit at a time so your quotient for each step must be between 0–9. If you get an answer larger than 9, you either divided the last step incorrectly or the error is in this step. If you can get zero groups, you need to add a zero to your quotient and bring down the next number.

$$28 \times 1 = 28$$
$$28 \times 2 = 56$$
$$28 \times 3 = 84$$
$$28 \times 4 = 112$$

And so on. But this can be time-consuming. We have another tool at our disposal. Estimation! 28×10 is 280. That's too big, but it is close to 268. Let's try 9. $9 \times 28 = 252$. That works!

$$
\begin{array}{r}
96 \\
28 \overline{)2{,}688} \\
-2\,52\downarrow \\
\hline
168 \\
-168 \\
\hline
0
\end{array}
$$

We subtract 252 from 268 = 16. Bring down the 8. Now we need to find how many groups of 28 are in 168. Let us estimate again. 5×28 is 140. That's too small, but it's close. What about 6? $6 \times 28 = 168$!

 HELPFUL HINT

Don't erase or scribble out your side multiplication. You may find it helpful later in the problem. Don't waste time doing the same multiplication twice!

Sometimes even when you repeat the process your number won't divide evenly. There will be pieces left over. You have written that before as a remainder.

For example, 1,517 ÷ 25 = 60 remainder 17. Do not be alarmed by large remainders. As long as the remainder is smaller than the divisor, it is correct.

Remainder as a Fraction

Remainders can be expressed as a fraction. The remainder is expressed over the divisor. The divisor is the denominator and the remainder is the numerator.

$$1,517 \div 25 = 60\frac{17}{25}$$

If you are curious about remainder as a decimal, we will explore that in the next section!

💡 HELPFUL HINT

The PARCC has said that this will be a 5 digit dividend and a 2 digit divisor and that it may or may not have a remainder.

Practice

1. 15,087 ÷ 47
2. 21,222 ÷ 81
3. 25,507 ÷ 26

Answers

1. **321**
2. **262**
3. **981 remainder 1** or **981$\frac{1}{26}$**

PARCC Question

A farm in Florida is trying to ship oranges to a grocery store in New Jersey. They have to get 10,527 oranges there as soon as possible. 40 oranges can fit in a crate, and crates take 17 hours by train to arrive. How many crates are needed to get all the oranges to New Jersey?

Answer Explained

We start by doing long division on 10,527 ÷ 40 and get 263 remainder 7. The word problem told us that all the oranges have to get there, so we need to add an extra crate for the remaining oranges. So 264 crates are needed. Keep in mind that 17 hours is a distractor piece of information.

Adding and Subtracting Decimals

Adding and subtracting decimals is simply testing your ability to "set up" a problem.

The important thing is place value. When you were younger and learning to add or subtract 2 digit numbers your teacher would remind you to "line up" your numbers carefully. That is what we need to do here. We need to line up our decimals.

$$
\begin{array}{ll}
37.15 & 412.36 \\
+4.2 & -25.264
\end{array}
$$

Notice that when we start with the decimal, all our place values are lined up (tens, ones, tenths, . . .).

 **HELPFUL HINT**

Do not be fooled if the numbers look messy. It is about digits! If you are having a really hard time, draw a place-value chart around the digits.

Feel free to add zeros into any "holes" if it helps keep your digits lined up. It is okay to add zeros because it doesn't change the value. For example, 0.3 is the same as 0.30 and they are read the same way, three tenths. In fractions, we know that $\frac{3}{10} = \frac{30}{100}$ because we multiplied both the numerator and the denominator by the same number (10). In decimals, that means that 0.3 and 0.30 are also equal, because 0.30 is $\frac{30}{100}$. When we add zeros, we are keeping the place value the same. We must add a zero when we are subtracting. It is important when we need to borrow.

For example,

$$
\begin{array}{ll}
\overset{1}{3}7.15 & \overset{3\ 10\ 12\ \ 2\ 15\ 10}{\cancel{4}\cancel{1}\cancel{2}.\cancel{3}\cancel{6}\cancel{0}} \\
+4.2 & -25.264 \\
\hline
41.35 & 387.096
\end{array}
$$

Practice

1. 133.215 + 12.3

2. 17.1 + 6.31

3. 84.7 − 8.92

4. 27.6 − 27.31

(handwritten work)

```
  133.215        17.1        84.7¹⁰
+  12.3        + 6.31      - 8.92
─────────      ───────     ───────
 145.515        23.41       75.78
```

```
  27.6⁵/¹⁰        133.215
-  27.31        + 12.3
─────────       ───────
  00.29           5.515
```

```
 133.215
+  12.3
────────
 145.515
```

Answers

1. **145.515**

2. **23.41**

3. **75.78**

4. **0.29**

PARCC Question

Carson is running a race that is 14.7 miles, and he has 3.27 miles left to go. Nell is running a race that is 12 miles, and has 1.4 miles left to go. Who has run a greater distance? How much farther did that person run?

Answer Explained

First we have to solve for Carson: 14.7 − 3.27 = 11.43.

Next we solve for Nell: 12 − 1.4 = 10.6.

Carson ran further.

To figure out the next step, we have to subtract their two distances

$$11.43 - 10.6 = 0.83$$

Carson has run 0.83 miles farther than Nell.

Multiplying and Dividing Decimals

When we multiply decimals, we line up the digits along the right side as we normally would when multiplying whole numbers. The easiest way to multiply a decimal is to multiply as if there is NOT a decimal there.

$$
\begin{array}{r}
153 \\
\times 768 \\
\hline
1{,}224 \\
9{,}180 \\
+107{,}100 \\
\hline
117{,}504
\end{array}
$$

$$
\begin{array}{r}
15.3 \\
\times 7.68 \\
\hline
\end{array}
$$

Now we have to put back in the decimal. Count the total number of decimal places from **both** factors. For example,

0.2 is one decimal place

0.13 is two decimal places

0.179 is three decimal places

We can relate this to fractions. For example, 0.2 is a denominator with 1 zero. When we divide by 10, we move place value to the left. 0.13 is a denominator with 2 zeros. When we divide by 100, we move 2 place values to the left. Finally, 0.179 is a denominator with 3 zeros. When we divide by 1,000, we move 3 place values to the left.

In our problem there is one decimal place in the top factor and two in the bottom. 1 + 2 = 3. My answer needs three decimal places. So 117,504 becomes 117.504.

Why does this work? What we are really doing is multiplying by 10 or 100 or 1,000 to get rid of the decimal and then dividing the same number at the end to put back in the decimal.

When we are dividing decimals, we write our long division problem as we have learned earlier. We complete long division following our usual rules. Be sure that in your long division you keep the quotient lined up correctly. After we have an answer, we carry the decimal up into the answer.

$$
\begin{array}{r}
1.9 \\
4 \overline{)\ 7.6} \\
-4 \\
\hline
3.6 \\
-3.6 \\
\hline
0
\end{array}
$$

Our answer is the GCF of 36 and 54 is 18.

If you know your series of primes, found with prime factorization, you can use this to find your GCF.

First we find our primes using a factor tree.

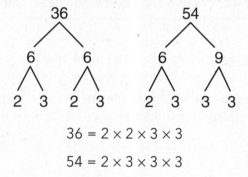

$$36 = 2 \times 2 \times 3 \times 3$$
$$54 = 2 \times 3 \times 3 \times 3$$

We look for any prime listed on both lists.

These two numbers have a $2 \times 3 \times 3$ in common, which gives a total of 18.

The Least Common Multiple (which we will refer from here on out as LCM) can be found by listing multiples of a number in order. Identify multiples that are on the lists of both numbers. The first (smallest) multiple on both lists is the LCM.

$$8 = 8, 16, 24, 32$$
$$12 = 12, 24$$

The first number on both lists is 24 (If we go further we would find that 48 is also on both lists. Remember that the object is to find the smallest number. Once you find it STOP!)

If we know the series of primes, we can use this to find the LCM.

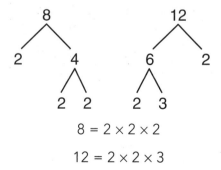

$$8 = 2 \times 2 \times 2$$
$$12 = 2 \times 2 \times 3$$

We need all the factors on both lists. We need $2 \times 2 \times 2 \times 3$ or 24. Any factor that overlaps counts once.

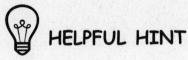 **HELPFUL HINT**

PARCC has said that you will need to be able to find the GCF of 2 numbers less than or equal to 100 and the LCM of 2 numbers less than or equal to 12. Focus your review on these numbers only!

We can use a Venn diagram to find both the LCM and the GCF. First, we find the primes using prime factorization. Anything shared goes in the middle section, and anything that they have independently is placed in the outer sections. To find the GCF, we multiply any number in the center section. To find the LCM, we multiply the entire diagram.

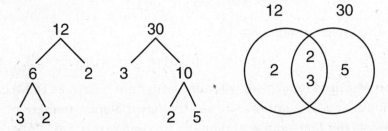

Inside Only 2 x 3 = 6 = GCF

Entire Diagram 2 x 2 x 3 x 5 = 60 = LCM

 HELPFUL HINT

In a word problem, often the **LCM** is a cycLe question and the **GCF** is a sharinG question.

Practice

Find the GCF.

1. 20 and 35 = 5

2. 28 and 56 = 28

3. 45 and 72 = 9

4. 12 and 82 = 2

Find the LCM.

5. 6 and 8 = 24

6. 4 and 12 = 12

7. 3 and 8 = 24

8. 9 and 11 = 99

Answers

1. **5**
2. **28**
3. **9**
4. **2**
5. **24**
6. **12**
7. **24**
8. **99**

PARCC Questions

1. 36 is the LCM for

$$\begin{array}{r} 12 \\ 3\overline{)36} \\ 3\downarrow \\ \hline 0\ 6 \end{array}$$

- ● A. 3 and 12
- ● B. 4 and 9 *Mental Math*
- ○ C. 9 and 12
- ○ D. Both B and C
- ○ E. A, B, and C
- ○ F. None

2. Tom is making care packages for people in nursing homes. He wants all the packages to be exactly the same. He wants to use all his materials and to make as many packages as possible. He has 21 pairs of fuzzy socks and 49 hard candies. How many care packages can Tom make? How many socks and how many candies are in each bag?

Answers Explained

1. **D** is the best answer. In A 3 and 12 do have a common multiple of 36, but they also have 12 as a common multiple, 12 is lower than 36. Both B and C are correct answers, making D the best answer.

2. This question is looking for a GCF. Tom can make 7 gifts that contain 3 pairs of socks and 7 hard candies each.

Negative Numbers

Negative numbers are numbers less than zero. They exist on the number line to the left of zero. They are also below zero on a vertical number line. These numbers are represented in a lot of real life situations. Negative numbers are used to describe elevations below sea level, temperatures below freezing, a bank account with debt, negative charges when studying electricity, and a display of how much time is left on a song or video.

Knowing that these numbers exist is the first step to being able to use them! The second is realizing that the larger the digit is farther away from zero, the less its value will be. For example −6 is less than −2. It is also important to know that fractions and decimals exist in between whole negative numbers the same as they do in the positive numbers.

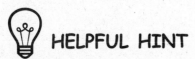 HELPFUL HINT

Drawing a quick number line will help prevent confusion when dealing with negative numbers.

Practice

Which of the following situations is a negative number?

1. The temperature at the polar ice caps

2. The elevation at the top of a mountain

3. How much of your movie you have already watched

4. The bottom of the ocean

5. Your body temperature

6. The amount of money in your piggy bank

Answers Explained

1. F. Temperatures get colder with larger negative numbers.

2. T. 2 is larger than all negative numbers.

3. T. It could have been any number smaller than –7. So –8, –9, –10, and so forth work.

4. F. –1 is bigger than –7.

5. F. –2 is smaller than 2.

Absolute Value

Absolute value is about distance from zero. If I have a ball in the middle of the field and Joey is standing 3 feet to the right and Billy is standing 3 feet to the left, they are both the same distance just in opposite directions. They are both 3 feet away. The same is true of the absolute value of 3 and –3. They are both three steps away from zero. One moves three spaces to the left and the other moves three spaces to the right on a number line.

Consider also that –7 is the opposite of 7. They have the same absolute value of 7, just in opposite directions. Absolute value is how far a number is away from zero. Both positive and negative numbers have an absolute value that is a positive number. The absolute value of 9 is 9, and the absolute value of –9 is also 9. Absolute value is shown when the number is between a pair of parallel vertical lines. |–6| is the absolute value of –6, which we now know is 6!

What is the absolute value of –2? |–2| = 2

 **HELPFUL HINT**

Remember absolute value is always positive. Many students make the error that the absolute value of a positive number is negative. It is positive, too!

Practice

Solve and complete the chart using <, >, =.

1. |7| |−13|

2. |2 × 6| |3 × 4|

3. |−12| |14|

4. |−11| |−14|

5. |−40| |5 × 4 × 2|

Answers

First, we simplify any expressions inside the absolute value. Then we remember that all numbers are positive after we solve the absolute value.

1. **7 < 13**

2. **12 = 12**

3. **12 < 14**

4. **11 < 14**

5. **40 = 40**

PARCC Question

Sara is studying the water level in a local river. She is basing her water measurements off of sea level. Sometimes the water level is above sea level, and other times it is below. Complete the chart below to show the readings as an absolute value.

Day	1	2	3	4	5	6	7
Measurement	0	4	3	−2	−4	6	1
Absolute value							
Above or below sea level							

Answers Explained

Day	1	2	3	4	5	6	7
Measurement	0	4	3	−2	−4	6	1
Absolute value	0	4	3	2	4	6	1
Above or below sea level	Sea level	Above	Above	Below	Below	Above	Above

All answers are positive, a distance from zero. Zero is the absolute value of zero. Negative numbers are below sea level, positive numbers are above sea level, and 0 is at sea level.

Ordering Integers and the Absolute Value of Integers

In the previous sections we learned how to order integers and absolute values. Now we need to be able to look at them together.

Let's look at the numbers 11, −11, |11|, |−11|. Which of these have the same value? Three of the numbers, 11, |11|, |−11| all equal 11. The only one number that is different and maintains a negative value is −11. It is also important to note what happens when the negative is on the outside of the absolute value symbol. For example: −|11| is equal to −11 and −|−11| is also equal to −11. The negative sign inside goes away but the negative on the outside remains.

| | Equal to 16 | Equal to −16 |
| | 16, |−16|, |16| | −16, −|−16|, −|16| |

 HELPFUL HINT

Absolute value symbols are grouping symbols. We solve what is inside before moving on.

Once we solve for the value of absolute value it is easy to decide which is larger. |2 × 6| is > 10 because 12 is greater than 10. |−16| > 3 × 2 because 16 is greater than 6.

Practice

Put the following numbers in order from least to greatest.

−2, 17,−|13|, −27, |−21|, |6|

Answers

−27, −|13|, −2, |6|, 17, |−21|

Because their value is −27, −13, −2, 6, 17, 21

PARCC Questions

Which of the following statements are false? Choose all that apply.

A. $17 < |-21|$

B. $-7 = -|7|$

C. $23 > |-23|$

D. $14 = |-14|$

E. $|-12| = |-12|$

Answers Explained

Choices A. and C. are both false. $-7 = |7|$ is really saying that $-7 = 7$ and it should be $-7 < 7$, and $23 > |-23|$ is really saying that $23 > 23$ and it should be equal.

Graphing in a Four-Quadrant Coordinate Plane

The coordinate plane is created by a vertical and a horizontal number line intersecting at 0. This creates 4 different quadrants (or quarters). These 4 quarters are known as Quadrant 1, Quadrant 2, Quadrant 3, and Quadrant 4, usually written as Quadrant I, Quadrant II, Quadrant III, and Quadrant IV.

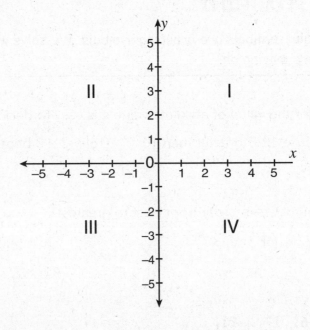

These 4 quadrants represent all positive and negative coordinate pairs. You have probably graphed before on Quadrant 1, which is made up of a positive x value (left and right) and a positive y value (up and down). The other 3 quadrants are created by adding in negative numbers. If the x is positive and the y is negative (+, −) we will graph in Quadrant 4. If both values are negative (−, −), we will graph in Quadrant 3. If the x is negative and the y is positive (−, +) the point will be located in Quadrant 2.

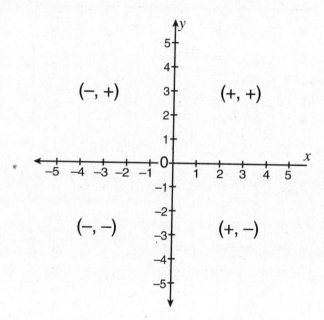

Remember, just like number lines, other numbers as well as fractions and decimals exist between labeled lines.

Plot and Name Points

In plotting points in the 4 quadrants, we use the same rules as we did when plotting in the single quadrant. The first number in the coordinate plane is x and the second is y. The first number decides the left or right movement. The second number decides the up or down movement.

Practice

Point A is (3, −2). From 0, move right to 3 and then down to −2.

Point B is (−3, −2). From 0, move left to −3 and then down to −2.

Point C is (−3, 2). From 0, move left to −3 and then up to 2.

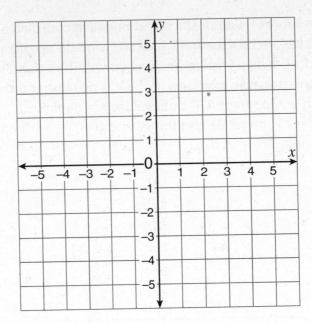

Answers

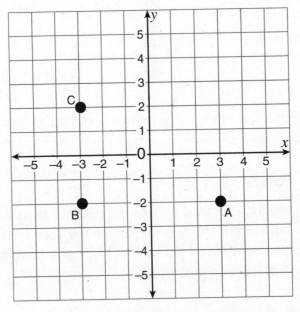

When two points have the same *x* coordinates, they form a **vertical** line. When two points have the same *y* coordinates, they form a **horizontal** line. You can easily find the distance between these two points on this line by subtracting the absolute value of the two numbers if they are either both positive or both negative. If they are different, one positive and one negative, we add the absolute values.

Find the difference between pints (3, 4) and (3, −7). These are different signs so add the absolute values |4| + |−7| =11.

Find the distance between points (5, 4) and (9, 4).

These are same signs so subtract the absolute values |9| − |5| = 4.

Find the distance between points (−3, −7) and (−8, −7).

These are same signs so subtract the absolute values |−8| − |−3|= 5.

Practice

1. (2, −7) and (6, −7)
2. (9, −7) and (9, 7)
3. (2, 7) and (−3, 7)

Answers

1. Same signs so we subtract the absolute values |6| − |2|= 4
2. Different signs so we add the absolute values |7| + |−7| = 14
3. Different signs so we add the absolute values |2| + |−3| = 5

PARCC Question

In Anthony's town the streets are laid out like a coordinate grid. The candy shop is located at (4, −2). The children have to walk exactly 6 blocks after school to get to the candy store. The children must stay on the sidewalk so they cannot cross blocks diagonally. If each unit grid is one block, what are possible locations for the candy store?

Location	Could represent a location of the candy store	Could NOT represent a location of the candy store
(−2, −8)		
(0, 0)		
(10, −2)		
(−4, 2)		
(−2, −2)		
(4, 8)		

Answers Explained

These locations can be the answer:

(0, 0) is 4 blocks down and 2 blocks right = 6.

(10, –2) is 6 blocks left.

(–2, –2) is 6 blocks down.

These locations can't be the answer:

(–2, –8) is 6 blocks in on each axis = 12 blocks.

(–4, 2) is just a reversal of signs and is, in fact, 12 blocks away.

(4, 8) is 10 blocks away. The mistake is with the negative sign (4, –8) is a correct answer.

Review Test

1. Abraham is filling up his sister's baby pool. The pool holds $6\frac{1}{2}$ gallons of water. He is using a bucket that holds $1\frac{1}{3}$ gallons. How many times will he have to fill the bucket?

2. Eric is organizing his workshop. He has 213 nails in each box. He has 34 boxes. How many nails does Eric have?

3. The sixth grade class is reviewing division of fractions. They know that they have to flip a fraction and then multiply. Dana says that you flip the first fraction. Marie says that you flip the second fraction, and Antonio tells them to stop arguing because they both will get you the same answer. Which student is correct?

4. Samantha is working on her addition of decimals. She keeps getting the wrong answer. Explain what she is doing incorrectly and find the correct answer.

$$\begin{array}{r} 26.1 \\ +1.20 \\ \hline 38.1 \end{array}$$

5. A fruit stand sells apples in 3.2 pound bags. The fruit stand owner received a shipment of 27.6 pounds of apples. How many bags can he make?

6. Maryanne wants each wall section of the art gallery to have the same combination of photos and paintings. She has 32 paintings and 56 photos in the show. What is the most wall sections she can fill? How many paintings and how many photos can go on each wall section?

7. The gym teachers have students sit in squads. Mr. Wilson has students sit in squads of 8. Ms. Lee has students sit in squads of 10. They both have the same number of students in their class. What is the least number of students they can have?

8. Mr. Berry's World History class is doing a final project. They are allowed to pick any year in history and research life during that time period. Below is a table of student sections.

Student 1	800 A.D.
Student 2	1420 A.D.
Student 3	2400 B.C.
Student 4	1300 B.C.
Student 5	1150 A.D.
Student 6	2100 B.C.
Student 7	421 A.D.
Student 8	120 B.C.

Write each student's year selection as an integer. Which integer is largest? Which integer is smallest? Using absolute value, what is the length of time that elapsed between the earliest point in time and the latest?

9. Solve the following

$$35.261 - 2.378$$

$$16.8 \times 4.345$$

$$184.6 \div 2.4$$

$$23.4 + 6.324$$

$$17.32 - 6.9$$

$$24.345 + 2.3481$$

$$28.14 \div 8$$

$$24.65 \times 34.2$$

10. Kristen says that 7 is larger than the absolute value of |–8| because 7 is larger than –8. Dominic says that |–8| is larger than 7 because 8 is more than 7. Who is right and explain to the person who is incorrect what he or she did wrong. If they are both incorrect explain both of their mistakes.

11. Todd graphs (3, –7) and (–2, –7) and connects the dots. Name and plot a point that will lie on his line.

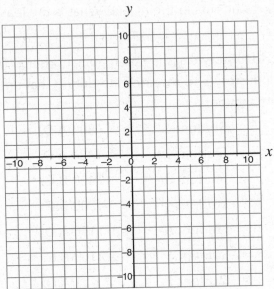

Answers

1. Division—First we make mixed numbers into improper fractions. $6\frac{1}{2}$ becomes $\frac{13}{2}$ and $1\frac{1}{3}$ becomes $\frac{4}{3}$. Then using $\frac{13}{2} \div \frac{4}{3}$ multiply the reciprocal $\frac{13}{2} \times \frac{3}{4} = \frac{39}{8}$ and make a mixed number $= 4\frac{7}{8}$. Thus, the answer is 5.

2. Simple multiplication $213 \times 34 = 7,242$ nails

3. Marie is correct. You will get different answers and we follow the rule Keep Change Flip! Keep (the first fraction), Change (the sign), and Flip (the second fraction).

4. Samantha is wrong because she lined up her digits incorrectly. We always line up along the decimal point. This gives her the correct answer of 27.3.

$$\begin{array}{r} 26.1 \\ +1.2 \\ \hline 27.3 \end{array}$$

5. Long division question—After we move the decimal in the divisor, we have the problem of 276 ÷ 32, which is 8.625. Then round up to 9 bags of apples.

6. Greatest Common Factor. They both have a factor of 8. So we have 8 wall sections with 4 paintings and 7 photos on each.

7. Least Common Multiple (LCM). The LCM of 8 and 10 is 40. So they each have 40 students.

8. 800, 1420, –2400, –1300, 1150, –2100, 421, –120

    The largest number is 1420 and the smallest is –2400. This is a difference of 3,820 years.

9.

    32.883

    72.996

    76.916

    29.724

    10.42

    26.6931

    3.5175

    843.03

10. Dominic is correct, telling Kristen that it is important to find the absolute value first before comparing. |–8| is 8 and 8 is larger than 7.

11. Any of the points with a y value of –7 and an x greater than –2 but less than 3.

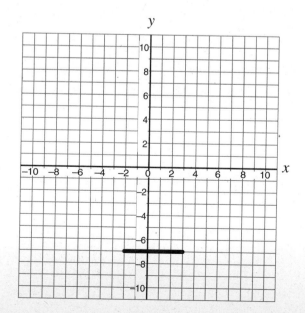

Ratios, Rates, and Proportional Relationships

Ratios, rates, and proportions all look at the relationships between numbers. **Ratios** are used when we are comparing quantities, or how many, or how much we have of something. They allow us to look closely at the relationship between numbers. The term "for every" is a common clue to a ratio. Ratios and rates are often discussed interchangeably, but typically a ratio is used to describe when everything has the same unit or label (cups, miles, ounces), and **rates** are used to compare things with different units (tablespoons per cup, miles per hour). **Proportions** look at scaling the numbers larger and smaller but keeping the same base ratio, or in other words finding equivalent ratios.

Two Quantity Comparisons

In looking at ratios, first we have to think about things we can compare:

The number of boys to girls in a classroom

The number of dogs to the number of cats at the pet store

The amount of oil to vinegar in a salad dressing

 HELPFUL HINT

Some ratios have expected answers. For example, for every dog, we have four legs, or for every one bicycle, we have two tires. Years have 12 months.

Ratios can be written in a number of ways

	With Labels	**Without** Labels
Using words	3 boys to 5 girls	3 to 5
Using colons	3 boys: 5 girls	3:5
As a fraction	$\dfrac{3\,boys}{5\,girls}$	$\dfrac{3}{5}$

These relationships can be **part to part comparisons** (3 boys to 5 girls) or they can compare a **part to a whole** (3 boys out of every 8 students).

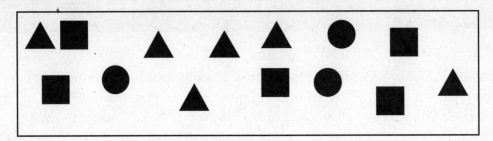

Practice

1. What is the ratio of triangles to squares?

2. What is the ratio of circles to triangles?

3. What is the ratio of squares to all shapes?

Answers

1. The ratio of triangles to squares is 6 to 5 because we count 6 triangles and we count 5 squares.

2. The ratio of circles to triangles is 3 to 6 because we count 3 circles and 6 triangles.

3. The ratio of squares to all shapes is 5 to 14 because we count 5 squares and there are 14 total shapes.

PARCC Question

Part A

Dillon has 9 coins in his pocket. He only has pennies and nickels. He has 3 more nickels than pennies. Write a ratio of pennies to nickels.

Part B

What if Dillon gave 2 pennies to a friend? How would that change the ratio?

Part C

How much money does Dillon have in part A? What is another ratio of pennies to nickels that would equal the same amount?

Answers

Part A

He has 3 pennies to 6 nickels or 3:6.

Part B

If he gives 2 pennies away he now has a ratio of 1:6.

Part C

$3 \times 1 = 3$ and $6 \times 5 = 30$ for $3 + 30 = 33$ cents total; other possible ratios of pennies to nickels that equal 33 cents are 8:5, 13:4, 18:3, 23:2, and 28:1.

Equivalent Ratios

Ratios are sometimes written as reduced to lowest terms as we do with fractions. Other times we need to expand the quantity. Creating **equivalent ratios** is the same process that we create equivalent fractions. We need to increase or decrease both numbers by the same **factor**.

If a class has 9 boys and 15 girls, we can reduce that ratio from 9:15 to 3:5. We reduce a ratio by finding a common factor. Both 9 and 15 are divisible by 3. This leaves us with 3:5.

A ratio of 6 stars to 9 hearts can be reduced to a 2:3 ratio. 6 divided by 3 is 2, and 9 divided by 3 is 3.

Practice

Reduce these ratios.

1. 150 students to 30 teachers

2. 24 gallons of iced tea for 16 gallons of lemonade

3. 15 tables for every 60 chairs

Answers

1. 5 students to 1 teacher

2. 3 gallons of iced tea for 2 gallons of lemonade

3. 1 table for every 4 chairs

We are sometimes asked either to determine if two given ratios are equivalent or to find matches. There are several ways to find the answer.

First, we can see if these two ratios have the same reduced ratio.

For example, if we reduce the ratio of daily quantities of 400 chocolate milks to 250 plain milks, we get 8:5. Similarly, if we reduce another school's ratio of daily quantities of 1,600 chocolate milks to 1,000 plain milks, we get 8:5. Both schools have the same ratio of students ordering chocolate milk; one is just a much larger school.

Reducing ratios is the exact same process as finding simplest form with fractions. Remember: We said that one way to write a ratio was as a fraction. When we find the simplest fraction we are also finding the lowest ratio!

$$36:6 \text{ reduced by } 6 \text{ is equal to } 6:1$$

$$\frac{36}{6} \text{ reduced by } \frac{6}{6} \text{ is equal to } \frac{6}{1}$$

We can also compare using the cross-multiplication method.

Write the two ratios as a fraction and cross-multiply. When we **cross-multiply**, we are multiplying the denominator of one fraction (in this case the fraction representing a ratio) and the numerator of the other. If the products are the same, we have equivalent ratios!

Are 75:40 and 90:48 equivalent?

 $75 \times 48 = 3,600 \quad 40 \times 90 = 3,600 \checkmark$

These ratios are equivalent.

💡 HELPFUL HINT

Use labels and be sure the same label is on the top of both fractions.

Practice

Find the ratio in each set that is NOT equivalent.

1. 42:36   25:18   21:18

2. 54:24   81:36   74:54

3. 22:34   66:102   77:109

Answers

1. **25:18**
2. **74:54**
3. **77:109**

 HELPFUL HINT

If in two ratios the first number is the same but the second one isn't, they are not equivalent. Also, if the second number is the same but the first one isn't, they are not equivalent. This is also true when we are comparing fractions. When we have two that have the same numerators but different denominators, the fractions cannot be equivalent. For example, 3:7 ≠ 3:9 and 5:7 ≠ 3:7.

PARCC Question

List 3 ratios that are equivalent to 2:7. Give one example of a ratio that is not equivalent. Explain why it isn't equivalent.

Answers Explained

Correct ratios include any ratio that multiplies the 2 and the 7 by the same factor. Common answers will include 4:14, 6:21, 20:70, 10:35,. . . A non-example would be any number that does not fit this rule. For example, 3:8 is not an equivalent ratio because we added 1 to each number, and 7:2 is not because we switched the number.

Proportions

Proportions is another way of saying equivalent fraction. When we find equivalent fractions it is really a proportion. Proportions are also used in geometry to talk about scale. **Scale** is increasing or decreasing size in a proportional way. For instance, if you are a fan of model cars, they are often created in scale. Every part of the car is a replica of the original in a smaller size. A common scale is 1:64, meaning all measurements of the real car are 64 times larger than the model.

In the last section we used cross-multiplication to check our work. When working with proportions we can use this same concept to find a missing value. When we multiply the numerator of the first fraction and the denominator of the second fraction (diagonally) we know that the two diagonals should be equal. If we have three parts, we can find the missing part using this concept.

If you are missing a numerator:

First multiply across the pair where you have both numbers. Here $2 \times 9 = 18$

Next, divide by the other denominator

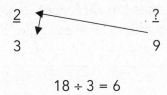

$$18 \div 3 = 6$$

The missing value is 6.

We can see that if we had scaled the ratio up by a value of 3 ($3 \times 3 = 9$, and $2 \times 3 = 6$) we would have gotten the same answer.

The same works for a missing denominator. Multiply across where you have both numbers and then divide by your other numerator.

Practice

Complete the following proportions.

$$\frac{10}{?} \quad \frac{40}{44}$$

$$\frac{8}{12} \quad \frac{?}{48}$$

$$\frac{?}{8} \quad \frac{21}{24}$$

$$\frac{2}{3} \quad \frac{18}{?}$$

Answers

$10 \times 44 = 440 \div 40 = 11$

$8 \times 48 = 384 \div 12 = 32$

$8 \times 21 = 168 \div 24 = 7$

$3 \times 18 = 54 \div 2 = 27$

PARCC Question

The local sports factory can make 1,200 footballs in 4 hours. How many can they make in 15 hours? Set up and solve with a proportion.

Answer Explained

To solve this problem we begin by turning our information into a proportion.

$$\frac{1,200}{4} \quad \frac{x}{15}$$

We put "x" on the top because that is the part we don't know (but a blank or a ? works just as well).

Now solve the proportion:

$$1,200 \times 15 = 18,000 \div 4 = 4,500$$

They will be able to make 4,500 footballs.

Unit Rates, Unit Pricing, and Constant Speed

A **unit ratio** is an equivalent ratio that uses the number 1. For example, if we start with a ratio of 25:5, we divide both numbers by 5 and get a ratio of 5:1. If I say that I need 1 cup of flour for every 3 cups of water that is a unit ratio. The 1 can be the first or the second term.

   Unit rate, a specific type of unit ratio, is a very important way to understand rates and be able to compare them. **Unit rates** are created when one of the numbers in the comparison is 1. We see unit rates all the time. Gas is sold per gallon. We buy fabric at a price per yard. We buy lunch meat per pound. Unit rates allows us to buy only the amount we need and compare which is the better deal. We use unit rates to find unit prices and constant speeds. The label of a unit rate is usually defined by "per," such as gallons per mile, dollars per hour, and gallons per minute. Whatever follows the per (mile, hour, minute) must have the value of 1; one mile, one hour, one minute, and so on.

Practice

What is the unit ratio for each of the following ratios?

1. 36:3

2. 7:42

3. 63:9

4. 18:4

Answers

1. **12:1**

2. **1:6**

3. **7:1**

4. **$4\frac{1}{2}$:1**

Unit pricing is a real-world application of unit rates. Grocery stores label each item with 2 prices. The first price is the price of the item, the second indicates a unit price. The unit price will tell us the cost per item, per ounce, per pound, and so forth. It allows us to easily see the value.

Practice

Break the following down into a unit rate.

1. Soda is sold in a case of 12 cans for $3.50. What is the price per can?

2. A box of markers has 8 markers and costs $1.20. What is the price per marker?

3. A 3 pound package of hamburger meat is $13.50. How much is it per pound?

Answers

1. **29 cents**

2. **15 cents**

3. **$4.50**

PARCC Question

We can use unit prices to find the best price. For example, a favorite brand of cereal comes in 3 different sized boxes. The 12 oz. box costs $3.75, the 20 oz. box costs $4.50, and the 35 oz. box costs $8.40. Understandably, the 35 oz. box costs more than the 12 oz. box, but is it the best value?

Answer Explained

When we divide by the number of ounces in each box, we find the cost per ounce. The 12 oz. box has a unit price of 31 cents per ounce; The 20 oz. box has a unit price of 22 cents per ounce, and the 35 oz. box has a unit price of 24 cents per ounce. The 20 oz. box has the lowest cost per ounce and is therefore the best value.

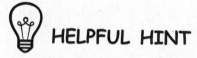 **HELPFUL HINT**

Don't be fooled into thinking that the largest size is always the best value despite what Value Clubs try to tell us!

Constant speed takes the total distance we travel and how long it takes us to get there to figure out how fast we are going. Again, we use the same model to solve as we did in unit rate. If I drive 250 miles and it takes me 5 hours, I can use division to figure out my miles per hour. 250 ÷ 5 = 50. I am driving an average speed of 50 miles per hour.

Practice

1. If Francisco's boat can travel 1,926 knots in 48 hours, how many knots per hour does he travel?

2. If Amed can run 17 miles in 3 hours, how many miles per hour can he run?

3. The speed limit on I-295 is 65 miles per hour. How long will it take you to drive 1,700 miles?

Answers

1. 1,926 ÷ 48 = 40.12 knots per hour

2. $17 \div 3 = 5\frac{2}{3}$ miles per hour

3. 1,700 ÷ 65 = 26.15 hours

PARCC Questions

1. Rachel is doing a report on different brands of cars. She really cares about the environment and wants one that is gas efficient. The three cars that she has chosen to research tell her different information about gas mileage. Car A has a sticker that tells her that it has a 22 gallon tank and can go an average of 629 miles per tank. Car B tells her that it can travel 1,000 miles in approximately 37 gallons of gas. Car C gets 30 miles per gallon. Which car has the best mileage?

2. Michelle is meeting her sister for vacation. Her sister has 1,216 miles to travel and is taking a train that travels an average of 80 miles per hour. Michelle is driving 630 miles at an average of 60 miles per hour. If they left at the same time, who will get there first?

Answers Explained

1. Car A gets 629 ÷ 22 = 28.6 mpg; Car B gets 1,000 ÷ 37= 27 mpg; and Car C gets 30 mpg. Car C has the best mileage.

2. Michelle is going to take 10.5 hours (630 ÷ 60). Her sister is traveling 1,216 ÷ 80 = 15.2 hours. Michelle is going to get there first.

Tables, Double Number Lines, and Tape Diagrams

Accurately comparing ratios requires organization. There are multiple ways to show your work for these kinds of problems. It is important that you know how to read and use these various **tools**.

Tables are one of the best and most common ways to organize information including ratios. It is very common that you will be asked to complete a table with equivalent fractions. You may see something like this:

3	9			72		540
7		49	70		1,050	

Completing this table requires us to ask ourselves what was multiplied to get from one ratio to another; and multiplying the other half of the chart by the same factor. For instance to get from the 3 to the 9 on the top line we multiply by 3. We therefore must multiply the 7 by the same factor resulting in the first blank box being filled with 3 × 7 = 21. We can continue completing the table remembering the number one rule: Whatever factor was used in the top must be used in the bottom.

The rest of the table is completed with the ratios 21:49 (a factor of 7), 30:70 (a factor of 10); 72:168 (a factor of 24); 450:1050 (a factor of 150); and 540:1260 (a factor of 180).

Practice

Complete this table that shows pet store purchases.

Goldfish	4	B	28	C	119
Guppies	A	18	63	117	D

Answers

A. **9**

B. **8**

C. **52**

D. **267.75**

Double number lines work much like a ratio table. However, unlike a ratio table that lists only certain specific ratios, double number lines show the continuum in relation to one another. We actually have examples of these in real life. **Rulers** frequently show both inches and centimeters. They are not split at the same rate.

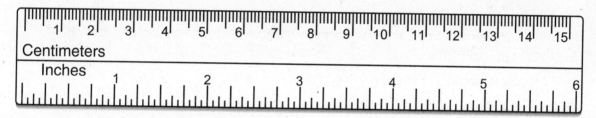

Many classic **thermometers** also show the temperature in both Celsius and Fahrenheit.

We can create a double number line to show any ratio relationship. We can create a number line that illustrates the average number of apples on apple trees.

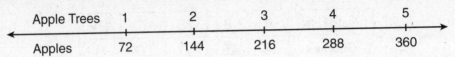

With this number line, we can see the whole number relationship but can also begin to see that $1\frac{1}{2}$ trees would yield approximately 108 apples. All ratios are represented.

Practice

Create a number line that shows the relationship of dog years to human years. Our starting ratio is one human year for every 7 dog years. About how many human years is the dog alive if he is 17 dog years old? Plot your answer on the line.

Answer

The dog is about $2\frac{1}{2}$ human years old.

Tape diagrams are another tool for modeling ratios that you may see on the PARCC.

Tape diagrams model an original ratio. We can use them to break down numbers. Let's say the perfect shade of purple is made from 3 parts blue and 2 parts red. I need a total of 40 gallons of paint. How many gallons of blue paint should I buy?

Let's start by building our tape diagram with original ratio.

Red

8	8

We have a total of 5 squares (2 red + 3 blue).

If I have 40 gallons total, I can divide 40 ÷ 5 = 8. Each square stands for 8 gallons. Write the number 8 in each square so that you can easily find the total of each color that I need.

Blue

8	8	8

We will need a total of 8 + 8 = 16 gallons of red.

We will need a total of 8 + 8 + 8 = 24 gallons of blue.

We can check our work 16 + 24 = 40. √

Problems can also be solved by setting up proportions and cross-multiplying. A proportion allows us to scale up or down an amount. If I know that I need to use 4 cups of water for every 3 cups of plaster, I can set up a proportion to figure out how much water 9 cups of plaster needs.

$$\frac{4}{3} = \frac{?}{9}$$

Now we need to cross-multiply, or multiply the numerator of the first fraction and the denominator of the other fraction, and then repeat with the other two numbers. The two cross products have to be equal.

$$4 \times 9 = 36 \qquad 3 \times ? = 36$$

$$3 \times 12 = 36$$

The answer is 12 cups of water.

Practice

Create a tape diagram to solve the following question:

Jamil has 7 baseball cards for every 2 cards his little brother Devonte has. Together the boys have 108 baseball cards. How many cards does each boy have?

Answer

Jamil

12	12	12	12	12	12	12

Devonte   Together there are 9 squares. 108 ÷ 9 = 12

Jamil has 12 × 7 = 84. Devonte has 12 × 2 = 24. 84 + 24 = 108√

12	12

PARCC Questions

1. A local animal rescue just received a huge influx of 325 dogs and cats. Adding the new animals made their ratio of dogs to cats 8:9. They already had 100 dogs and cats. How many dogs do they have now? Use any of the models to illustrate and explain your answer. Why did you choose the model you did?

Onions	1	2	3	4	5	6	7	8	9	10	11	12
Celery	3	6	9	12	15	18	21	24	27	30	33	36

2. Look at the double number line above.
   A. How many stalks of celery do you need per onion?
   B. How many stalks of celery would you need for 12 onions?
   C. How many onions would you need for 12 stalks of celery?
   D. If the ratio of carrots to celery was 2:1, how many onions would you need if you wanted to use 14 carrots?

Answers Explained

1. Models and reasons for the models will vary. We should have found a way to illustrate the 8:9 ratio increased to a factor of 25 to be 200:225, which gives us a total of 425 animals. And we should clearly indicate the answer of 200 dogs.

2. A. 3 stalks per onion
   B. 36 stalks of celery
   C. 4 onions
   D. There are 14 carrots for every 7 celery stalks. This means that for every 7 celery stalks, there are $2\frac{1}{3}$ onions because 7 divided by 3 is $2\frac{1}{3}$.

Graphing Rates and Proportions

When working with a large series of proportional relationships, it often helps to have a visual model. The best way we can do this is with a graph!

A recipe for the perfect party mix requires 16 pretzels for every 12 chocolate candies. If we are only making a snack for one person, it will be very different amounts than if we are making a snack for 20 people. We can show the various possible quantities on a graph!

Pretzels	Chocolate
60	45
140	105
220	165
300	225

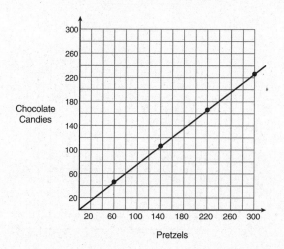

By using this graph we can look at any point on the line and it will give us a different set of numbers, and each one represents the exact same ratio of salty to sweet! We can eat 4 pretzels and 3 chocolates or we can serve an enormous bowl of 2,400 pretzels and 1,800 chocolate candies. Everyone will still taste the same perfection!

In our example the chocolate was the *x* axis and the pretzels were the *y*. The line represents the perfect combination. On the rest of the graph (but off the line) there are infinite combinations that are not the perfect ratio: too many pretzels for too little chocolate or the reverse. We can graph other things the same way. This can be useful for rates as well.

If you are working in a factory and can assemble 10 boxes in $\frac{1}{2}$ hour, we can create a line graph that shows expected productivity. Just like with a graph earlier, we start with a table.

Boxes	Hours
10	$\frac{1}{2}$
40	2
80	4
100	5

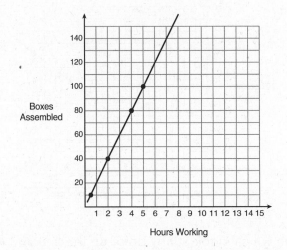

This line now represents expected productivity. If you work for 7 hours tomorrow, the boss can expect that you will assemble about 140 boxes.

Practice

Create a graph for a car that drives 45 miles per hour.

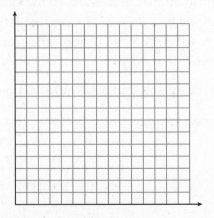

Answer

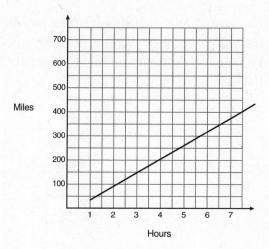

Graphs may differ depending on selected scales.

PARCC Question

Matilda found the following graph that showed the cost at a local salad bar. Have her write four true statements.

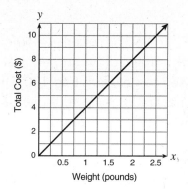

Answer Explained

There are infinite things that Matilda can write. It should come back to the unit price of $4 a pound. A good idea here would be to complete a table that shows possible cost scenarios. Also we can say things like it costs more than $3 a pound or less than $5. We can say that the graph only goes up to 2.5 pounds.

Percent

Percent is a very specific ratio. It relates the part to the whole. Percent is always "out of 100." 100% is 100/100 meaning 1 whole. For example, 70% is 70 out of 100 and 35 out of 100 is equal to 35%. If we think of percent as a ratio, it is easy to see how we calculate the percent of a number. We are simply making equivalent ratios!

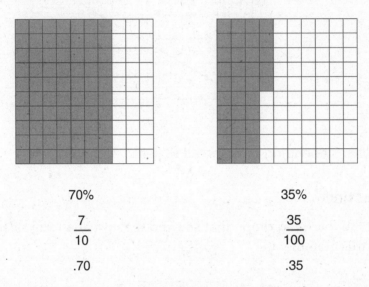

70%	35%
$\dfrac{7}{10}$	$\dfrac{35}{100}$
.70	.35

Say the original ratio is 3:5. To find a percent, we know we are trying to make a ratio of __ out of 100. Using our rules of finding equivalent ratios, we have to figure out what factor was used to get from 5 to 100. The answer is 20. By measuring 3 × 20, we get a new ratio of 60/100—that's 60%! Using the rules of equivalent ratios, we can now find the percent.

$$\frac{\text{Part}}{\text{Whole}} \quad \times \text{the same factor} \quad \frac{\text{Percent}}{100}$$

We can also set up a multiplication problem. The word "of" in percent problems means to multiply. 32% of 13 means 32% × 13. When working with percent, remember that percent is out of 100. $32\% = \dfrac{32}{100} = 0.32$.

There are two main ways to work with percent. The first is to solve rewriting percent as either a decimal (32% becomes .32) or by writing it as a fraction $\dfrac{32}{100}$. We can now use these numbers to solve the multiplication.

$$\begin{array}{r} 13 \\ \times .32 \\ \hline 26 \\ +390 \\ \hline 416 \end{array} \qquad \frac{32}{100} \times \frac{13}{1} = \frac{416}{100}$$

Add in decimal 4.16   Mixed fraction $4\frac{16}{100}$ (or $4\frac{4}{25}$)

The other way to solve is to set up a proportion.

$$\frac{Part}{Whole} \quad \frac{Percent}{100}$$

Remember we said that proportions follow the rules that diagonal products have the same value. The best part of proportions is that it doesn't matter what "pieces" you have and which piece you are missing, the problem is set up the same way.

What is 14% of 27?

$$\frac{Part}{27} \quad \frac{14}{100}$$

$$14 \times 27 = 378, \quad 378 \div 100 = 3.78$$

Brittany has a reading test with 35 questions. If she has 26 correct answers, what is her percent score?

$$\frac{26}{35} \quad \frac{Percent}{100}$$

$$26 \times 100 = 2600 \div 35 = 74.3\%$$

Practice

1. What is 75% of 130?

2. What is 22% of 625?

3. If I have 11 girls and 16 boys in my class, what percent of my class is girls?

4. If I have spent 13%, or $65, of my savings on a new video game, how much money did I originally have?

Answers

1. **97.5**

2. **137.5**

3. **40.7%**

4. **$500**

PARCC Questions

Part A

Felicia is shopping for new school clothes. She buys 2 sweaters, 3 pants, and a new dress. Her bill comes to $138. She has a coupon for 15% off. How much is her new total before tax? How much more would she save if she had a 20% coupon?

Part B

Christine likes to leave a 20% tip for her waiter. The tip that she left was $7. What was the new total of the bill?

Answers Explained

Part A

$117.30; $6.90. First, we have to ignore the distractor information of how many pieces of clothing she bought. Now we need to find 15% of 138, we multiply 0.15 × 138 which is $20.70. We need to be mindful that this amount represents the SAVINGS, not the new total. We need to subtract 138 − 20.70 to find the new total of $117.30. We repeat the process for 20% and find that 20% is a savings of $27.60. This savings is $6.90 larger than the savings she had at 15%.

Part B

Christine's original bill was $35. Using a Tape Diagram, we can show

7 dollars	7 dollars	7 dollars	7 dollars	7 dollars
20%	20%	20%	20%	20%

If we add her $7 tip to her $35 bill, we will find that she spent a total of $42.

Converting Measurement

When we are measuring something, it is important to know that there are different **units** or labels of measurement that can describe the same thing. For example, we can measure a sports field in meters or yards. We can measure out an ingredient in a recipe in either teaspoons or tablespoons. However, we also have to be careful that we do not think that one teaspoon is the **same** as one tablespoon!

If we know the conversion of two measurements, we can treat them like equivalent ratios too!

Let's start with something simple. One foot is equal to 12 inches also known as a 1 foot to 12 inch ratio. If we have 3 feet, we multiply by a factor of 3, so the new ratio is 3 feet: 36 inches. This tells us that there are 36 inches in 3 feet.

We can follow the same steps to make the conversion of cups to gallons. There are 16 cups to a gallon. If we have 500 cups of water, how many gallons do we have?

$$\frac{500\,\text{cups}}{?} \qquad \frac{16\,\text{cups}}{1\,\text{gallon}}$$

We don't know the number of gallons to complete this ratio. But if we use proportions, we know that 500×1 is 500, and then we need to divide that by 16. There are 31.25 or $31\frac{1}{4}$ gallons.

You can draw a picture to help you solve these problems. If you are trying to figure out the number of teaspoons in 8 tablespoons, you can draw a diagram to help.

$$\boxed{\text{Tablespoon}} = \begin{array}{l} \boxed{\text{Teaspoon}} \\ \boxed{\text{Teaspoon}} \\ \boxed{\text{Teaspoon}} \end{array}$$

Since there are 3 teaspoons in 1 tablespoon, we multiply 8×3 and get 24.

Practice

Use the conversion chart on the tools page at the back of the book to complete.

1. How many quarts are in 5 gallons?

2. How many inches are in 6 yards?

3. How many miles is 12,000 feet?

Answers

1. **20 quarts**
2. **216 inches**
3. **2.27 miles**

PARCC Question

Phillip, Francis, and Paul are competing to see who ran the farthest distance. Phillip ran 2 miles, Francis ran 2,000 yards, and Paul ran 10,000 feet. Which boy wins? How much farther would the other two boys have to run to catch up to the winner?

Answer Explained

In order to compare these three boys, we first have to make their distances all the same unit. If we make them all feet, Phillip ran 2 × 5,280 = 10,560 feet, Francis ran 2,000 × 3 = 6,000, feet, and Paul ran 10,000 feet. Phillip ran the farthest. Francis needs to run 4,560 feet or 1,520 yards, and Paul needs to run 560 more feet.

 HELPFUL HINT

You are not expected to memorize any conversions; they will be provided. You should know some basics like how many days, weeks, and months are in a year (365, 52, 12); how many inches are in a foot (12); and how many centimeters are in a meter (100). DO NOT FORGET TO KEEP LABELS ON YOUR ANSWER.

Review Test

1. Sharon is baking Christmas cookies. She bakes 7 dozen chocolate chip, 6 dozen snickerdoodles, 6 dozen cut-out sugar cookies, 4 dozen almond snowballs, 2 dozen jelly rings, and 5 dozen peanut butter blossoms.

    What is the ratio of snickerdoodles to snowballs?

    What is the ratio of jelly rings to sugar cookies?

    What is the ratio of peanut butter blossoms to all cookies?

    What is the ratio of chocolate chip and snickerdoodles to jelly rings and peanut butter?

2. Reduce the following ratios to lowest terms.

14:35

16:12

27:63

6:36

3. There are 121 vehicles in the parking lot. The ratio of cars to trucks is 7:4. How many trucks are in the lot?

4. The sum of two numbers is 169. The ratio of the same two numbers is 4:9. Find the bigger number.

5. Last week Margo bought 4 new nail polish colors and it cost her $12. Today she has only $9. How many new colors can she buy? Solve by setting up a proportion.

6. A farm has picked 36 pounds of peaches in the past 4 days. What is their average daily rate?

7. Ostriches have a top speed of 38 miles per hour. If they were able to travel at their top speed for one and a half hours, how far would they get?

8. It costs $2.56 for a dozen eggs. What is the unit cost per egg? (Rounded to the nearest penny.)

9. A 14.5 ounce can of tomatoes costs $0.89. A 20 ounce can costs $1.45. Which can is the better buy?

10. Complete the table of equivalent ratios.

4		32	8		48
	25	40		30	

11. Brenda's car travels 350 miles on a 14 gallon tank of gas. What is her average miles per gallon? Graph her average gas usage for the tank.

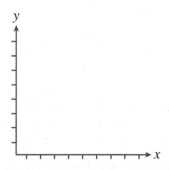

12. Carl pitched in the baseball game last night. He pitched 135 times and it was a strike 60% of the time. How many strikes did he pitch?

13. Mark lives in Florida 5 months out of the year and the rest of the time in New Jersey. What percent of the year (rounded to the nearest hundredth) does Mark live in New Jersey?

14. What is 74% of 550?

15. How many pints are in 3 gallons?

16. How many centimeters is 3 yards, if 1 inch is equal to 2.54 centimeters?

Answers

1. 6:4 or 3:2

   2:6 or 1:3

   5:30 or 1:6

   13:7

2. 2:5 Reduced by 7

   4:3 Reduced by 4

   3:7 Reduced by 9

   1:6 Reduced by 6

3. 44; the ratio shows that there are 11 total vehicles for each group of 7 cars and 4 trucks. There are 11 groups of 11 meaning that there are 77 cars and 44 trucks.

4. 117. You need to solve 9×13.

5. $\dfrac{4}{12} \quad \dfrac{?}{9}$ $4 \times 9 = 36$ $36 \div 12 = 3$. She can buy 3 colors.

6. $36 \div 4 = 9$ peaches per day

7. $38 \times 1.5 = 57$ miles

8. 21 cents per egg

9. The first can is 6 cents and the second can is 7 cents per ounce. The 14.5 ounce can is a better buy.

10.

4	20	32	8	24	48
5	25	40	10	30	60

11. 25 miles per gallon

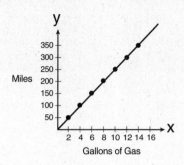

12. 81 strikes

13. He spends 58.33%—or 7 months—of his year in New Jersey.

14. 407

15. 24 pints. There are 8 pints to a gallon. $8 \times 3 = 24$

16. 274.32 centimeters. 3 yards = 9 ft = 108 in = 274.32 cm

Expressions and Equations

In this chapter, we are going to work on your algebraic thinking. We are trying to solve problems that have a missing piece. In algebra, we are working on writing and solving expressions, equations, and inequalities using a variety of different strategies. **Expressions** are the problems or formulas. They become an **equation** once we add the other side with an equal sign (=), or an **inequality** once we add the other side with an inequality symbol (<, >). We see equations and expressions in real-life situations with the use of formulas. **Formulas** are mathematical equations that are rules to solving a problem. For example, when you learned that multiplying the length and the width finds the area of a rectangle, that was a formula.

Exponents

When you first learned about multiplication, you were exposed to it as repeated addition. You knew how to add 2 + 2 + 2 + 2 and that it equaled 8. You then learned that the better way to write this was 2 × 4 because we had 4 groups of 2.

Exponents are the same thing: **exponents** are repeated multiplication instead of repeated addition.

There are two parts to an exponent. The number that is written larger and on the line is called the **base**. The base tells us what number we are multiplying. The number written smaller and to the upper right of the base is called the **power**. The power tells us how many times we will be multiplying the base. Together they make an exponent.

$$Base^{Power}$$

When you write out an exponent as a repeated multiplication it is called **expanding**. Unless you are solving exponents on a calculator, you will need to expand before you can solve.

$$5^3 = 5 \times 5 \times 5 = 125$$

$$6^2 = 6 \times 6 = 36$$

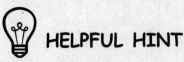

HELPFUL HINT

Don't confuse the base and the power. 5^3 means $5 \times 5 \times 5 = 125$ and is not the same as $3 \times 3 \times 3 \times 3 \times 3 = 243$.

If the power is outside parentheses, you expand the entire parentheses. Remember order of operations tells us to solve inside parentheses before we solve exponents.

$$(x + 2)^2 = (x + 2) \times (x + 2)$$
$$(y - 6)^4 = (y - 6) \times (y - 6) \times (y - 6) \times (y - 6)$$
$$(7 - 2)^2 = 5^2 = 25$$

The same rules apply if we are working with letters and symbols. For example,

$$d^4 = d \times d \times d \times d$$
$$f^6 = f \times f \times f \times f \times f \times f$$

When reading a power of 2, (5^2) is read as 5 to the 2nd power or 5 squared. When reading a power of 3, (5^3) is read as 5 to the 3rd power or 5 cubed. Beyond 3 we just call them by their level (4th, 5th, and so on).

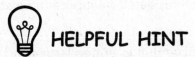

HELPFUL HINT

You will often see exponents with variables when you are dealing with squares. Area = s^2. Surface area of a cube = $6s^2$.

Practice

Expand and evaluate (if possible).

1. 6^3
2. 5^4
3. y^3
4. $(7 - 2)^2$

Answers

1. **6 × 6 × 6 = 216**
2. **5 × 5 × 5 × 5 = 625**
3. **y × y × y cannot solve**
4. **(7 − 2) × (7 − 2) = 5 × 5 = 25**

PARCC Questions

1. 3.2^3
2. 5.7^2
3. $\left(\dfrac{2}{5}\right)^3$
4. $\left(\dfrac{1}{7}\right)^4$

Answers Explained

1. **3.2^3 = 3.2 × 3.2 × 3.2 = 32.768**
2. **5.7^2 = 5.7 × 5.7 = 32.49**
3. $\left(\dfrac{2}{5}\right)^3 = \dfrac{2}{5} \times \dfrac{2}{5} \times \dfrac{2}{5} = \dfrac{8}{125}$
4. $\left(\dfrac{1}{7}\right)^4 = \dfrac{1}{7} \times \dfrac{1}{7} \times \dfrac{1}{7} \times \dfrac{1}{7} = \dfrac{1}{2,401}$

Variables

There are times that we are working with problems that have an unknown amount. You have seen this idea since you were very small. When you are asked for different ways to make 10, it was introducing this idea. The partner for 7 was 3. When you got a little older, this was shown as a blank. For example, 3 + __ = 7. You knew that the answer was 4.

Now instead of a blank we are going to use a variable. A **variable** is a letter or symbol that is used in place of an unknown number. A variable can change, or vary, which gives us the term *variable*. We can show our fill in the blank problem with a variable. $3 + n = 7$. We can easily see that $n = 4$. It doesn't matter if we say $3 + x$, or $3 + n$, or $3 + s$. The variable still equals 4.

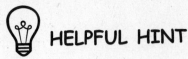

HELPFUL HINT

Be aware that variables are usually shown as *italics*.

Obviously the problems you will deal with in this chapter will not be so simple. First, we need to understand the vocabulary. **Terms** are the parts of an expression. Each term is a part of the problem or formula that is separated by either an addition (+) or subtraction (−) symbol. You will sometimes be asked to list or count terms.

In the expression $3x + 3y - 7$, there are three terms $3x$, $3y$, and 7.

In addition to the variables, there are numbers. These numbers can come two different ways. When they are used to describe the variable such as $3x$, they are called **coefficients**. If you remember that Co means "with" it is easy to remember that coefficients are with a variable attached. Coefficients tell us what to multiply by the variable. $5y$ means $5 \times y$. There are 5 ys. Another number found in an expression is a constant. **Constants** are terms that do not contain a variable; they are separated from variables by addition or subtraction. For example, in the expression $3 + x$, the number 3 is a constant. They are considered "constant" because the variable does not change their value.

Practice

Complete the table.

	Variables	Coefficients	Constants	Number of terms
$3x + 2$				
$\frac{1}{3}x - 11$				
$2x + 7 - y$				
$9n$				
$b + 7$				

Answers

	Variables	Coefficients	Constants	Number of terms
$3x + 2$	x	3	2	2
$\frac{1}{3}x - 11$	x	$\frac{1}{3}$	11	2
$2x + 7 - y$	x and y	2 and 1	7	3
$9n$	n	9	None	1
$b + 7$	b	1	7	2

PARCC Questions

Sort the following expressions under the correct heading.

    2 is a constant   2 is a coefficient

$2x + 3$

$3x + 2$

$2 + 3x$

$3 + 2x$

$2x + 2$

$3x^2 + 7$

Answers Explained

2 is a constant	2 is a coefficient
$3x + 2$	$2x + 3$
$2 + 3x$	$3 + 2x$
$2x + 2$	$2x + 2$

Note $2x + 2$ has both a constant and coefficient of 2. And the final expression is on neither list because an exponent of 2 is neither a constant nor a coefficient.

Writing Expressions and Equations

Often the first step to solving an expression or equation is to translate the words. This may be a word problem or simply the translation of a sentence. The secret here is to be able to decode the key words.

Words that clue addition = sum, also, added, addition, combine, together, perimeter

Words that clue subtraction = difference, increased, decreased, change, more, less

Words that clue multiplication = times, total, product, each, in all, area, factor, multiple

Words that clue division = quotient, equal parts, per, ratio, percent, split, shared

These key words are combined with numbers and variables in various ways to describe expressions.

Consider, for example, these expressions:

6 less than x

The product of y and 12

The combination of two factors $\frac{1}{3}$ and z

These phrases create the following expressions:

$x - 6$

$12y$

$\frac{1}{3}z$

When we are trying to write equations from phrases, we simply add words that clue an equal sign and represent the other side of the equation.

Words that clue equals = is, the same as, equals, equivalent, is balanced with

The sum of $3\frac{1}{4}$ and z is equivalent to $4\frac{1}{2}$ $3\frac{1}{4} + z = 4\frac{1}{2}$

36 is the same as the product of 7 and y $36 = 7y$

6 split into f groups is $\frac{2}{3}$ $6 \div f = \frac{2}{3}$

Many common real-world situations can be described as an equation. A car has four tires: $4 \times c = t$. Birds have two wings: $b \times 2 = w$.

 HELPFUL HINT

Be extra careful with the order of subtraction. 6 less than x means $x - 6$, unlike 6 minus x, which is $6 - x$.

On the test, you may be given a phrase and asked to write or match the correct expression or equation. It may also be written in the context of a more elaborate "story problem." The rules are the same. Find the important information, match the numbers and the key words, and decide what operation or operations are required of you.

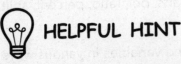 **HELPFUL HINT**

There is a highlight feature on most problems. It is a good idea to use this to highlight your clue words!

Practice

Write the following expression or equation.

1. A number x less than 8

2. The product of 6 and y is 30

3. $\frac{1}{2}$ times the value of x plus 2 is equal to 14

4. 16 less than a number z is 41

5. 9 groups of q is 27

Answers

1. $8 - x$

2. $6 \times y = 30$ or $6y = 30$

3. $\frac{1}{2}x + 2 = 14$

4. $z - 16 = 41$

5. $9 \times q = 27$ or $9q = 27$

PARCC Question

Jennifer is traveling to Italy on vacation. The maximum weight of luggage per passenger is w pounds. She has decided to take two bags, and her first suitcase weighs 27 pounds. Write an expression for the maximum weight, in pounds, her second suitcase could weigh.

Answer Explained

She can have a total of w, so we need to take away what was already used.

Our expression is $w - 27$.

Using Properties to Make Equivalent Expressions

In Chapter 2, we reviewed the Identity, Associative, Zero, and Commutative Properties. These properties allow us to rearrange problems in order to make them easier to solve. This is particularly handy when dealing with variables! The rules allow us to create equivalent expressions. **Equivalent Expressions** are expressions written in two different ways and have the same value.

The **Identity Property** tells us that any number times 1 is equal to itself. With variables, whenever we have a standalone variable, it is showing that the coefficient is 1. For example, $x = 1x$. We don't need the one in front.

$$5 \times x = (5 \times 1)x = 5x$$

The **Zero Property** is wonderful when we are working with variables. If we multiply any variable times zero, just like any number, it equals zero! Solving this problem is easy!

$$0 \times y = 0$$

$$5 + 0 \times f = 5 + 0 = 5$$

The **Commutative Property** is very important for combining **like terms**, or terms that are either constants or the same variable. This allows us to simplify confusing expressions. If we have $5 + x + x + 5 + x + 2$, we can move our terms around so that our numbers, our x, and our constants are together. $5 + 5 + 2 + x + x + x$. After grouping the like terms, we can see that some of this can be simplified. $5 + 5 + 2 = 12$, and $x + x + x = 3x$. Suddenly $12 + 3x$ seems a lot less messy.

The **Associative Property** also allows us to manipulate expressions. For example, let's multiply $3 \times 5y$. What that expression really says is $3 \times (5 \times y)$. The associative property tells us that we can group multiplication in a different way or $(3 \times 5) \times y$, and that means $15 \times y$ or $15y$ is the answer.

We also have a new property: The **Distributive Property**. The Distributive Property is really about order of operations. If we have to solve $3 \times (2 + 5)$, order of operations tells us that we must add $2 + 5 = 7$ and then we can multiply by 3 and get 21. If we disregard the parentheses, we get an incorrect answer of 11. The Distributive Property tells us that there is a better way.

Instead, we can distribute the 3. This means we multiply each piece separately by 3 and then add them together. $3 \times 2 = 6$ and $3 \times 5 = 15$ and $6 + 15 = 21$, which is the correct answer. This model may help this make more sense.

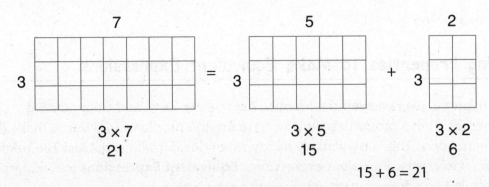

This property can help us multiply expressions with variables.

If we are multiplying $\qquad$ 3 (2x + 6)

We multiply each piece by 3 $\qquad$ 3 (2x) + 3 (6) = 6x + 18

The Distributive Property allows us to evaluate and simplify expressions without first solving for what is inside parentheses.

The Distributive Property can work the opposite way. When we go the opposite way, it is called **factoring.** Factoring is finding and separating factors. In this case, we are finding common factors.

If we are adding $\qquad$ 45 + 27

We can express them as $\qquad$ (9 × 5) + (9 × 3)

We can then factor out the 9 $\qquad$ 9 × (5 + 3)

No matter which of these three problems we solve, we get an answer of 72.

When the problem is written with the parentheses such as 3(2 + 6) it is called **factored form**. When we write it out as 3 × 2 + 3 × 6, it is called **expanded form.**

 HELPFUL HINT

If you start out adding, your answer should still have addition in your parentheses; the same is true if it is subtraction.

You may feel that this is unnecessarily complicated, and for the preceding problem, you would probably be correct! We are using that so we can understand the process better. This is useful if we have variables.

What if you were asked to add 4x + 32? You can't simply add because one term has a variable and the other does not. But we can factor it!

4x + 32

4(x) + 4(8) $\qquad$ Both terms have a factor of 4!

4(x + 8) $\qquad$ We have factored the expression and found an equivalent expression.

4x + 32 is the same as 4(x + 8).

HELPFUL HINT

Factoring and distributing are often used in word problems about the area of a rectangle. For example, if I told you that a carpet has an area of $6x - 18$ square feet and a width of $2x - 6$ feet, we can factor and see that $6x - 18 = 3(2x - 6)$. This means that the length of the carpet is 3 feet.

In all of these properties, the goal is the same: to find other ways to write the same expression. When we can change around the terms, we can sometimes find new information about an expression and use it to solve different problems. On the PARCC test, you will be asked in lots of different ways if you can identify and create expressions that are equivalent.

Practice

Write an equivalent expression using the named property.

1. $5 + x + 3$ Commutative

2. $2(3 + y)$ Distributive

3. $5 \times (6 \times f)$ Associative

4. $5x - 15$ Distributive

Answers

1. **$x + 5 + 3$ or $3 + 5 + x$ (or another arrangement)**
2. **$(2 \times 3) + (2 \times y)$**
3. **$(5 \times 6) \times f$**
4. **$5(x - 3)$**

PARCC Question

I want to find the perimeter of a square. One side measures $3x + 7$. Express the perimeter in only two terms.

Answer Explained

Perimeter is $s + s + s + s$ or $4(s)$. In this case, $4(3x + 7)$, which distributes to $12x + 28$, which is two terms.

Solving Expressions and Equations

One way to look at solving an expression or equation is to try to decide "does this fit?" In math, we call this substitution. **Substitution** is putting in a possible value for a variable and checking if that answer works, or inserting different values and finding the possible answers.

In expressions, we can find solutions when we are given a value to substitute in for the variable, for example, $x = 6$.

$6 + x$ becomes $6 + 6 = 12$

$x \div 3$ becomes $6 \div 3 = 2$

In equations, we are given a variable and asked if it COULD be the answer. For example,

$x = 6$

$6 + x =$ 12	$x \div 3 =$ 18
$6 + 6 =$ 12	$6 \div 3 =$ 18
$12 =$ 12 YES	$2 =$ 18 NO

In working with equations, we can also use the rules for **inverse operations**. Inverse means opposite, so inverse operations are opposite operations. Addition and subtraction are inverse operations to each other, and multiplication and division are inverse operations.

Think about it this way. If we have 6 and add 3, we get 9. If we subtract 3, we are back to 6. Adding 3 and subtracting 3 cancel each other out. The same is true if we multiply and then divide. By canceling out the numbers that are around the variable, we get the variable **isolated**, or alone, and can determine its value. Just remember that in order to keep the equation balanced, whatever we do to the left side we must also do to the right and vice versa.

$$
\begin{array}{llll}
w + 6 = 14 & x - 7 = 20 & 3z = 15 & m \div 12 = 9 \\
\underline{-6 \quad -6} & \underline{+7 \quad +7} & \underline{\div 3 \quad \div 3} & \underline{\times 12 \quad \times 12} \\
w = 8 & x = 27 & z = 5 & m = 108
\end{array}
$$

Sometimes a problem needs to be solved in two steps. When working to isolate a variable we use *reverse* order of operations. This means we need to do our adding or subtracting *before* we multiply or divide. That is because we can move the constant (the part attached by adding or subtracting) more easily than we can the coefficient (the part attached by multiplying or dividing).

$3x + 12 = 33$

First we move the 12 using inverse operations $3x = 21$

Now we move the 3 using inverse operations $\dfrac{3x}{3} = \dfrac{21}{3}$

We see that $x = 7$

When solving an equation using inverse operations, the last step is always to check with substitution. Take the solution and plug it back into the original equation to be sure that the answer checks out. If it doesn't, we can be sure there is a mistake!

$8 + 6 = 14 \checkmark$ $27 - 7 = 20 \checkmark$ $3 \times 5 = 15 \checkmark$ $108 \div 12 = 9 \checkmark$

When we are working with cross-multiplying and proportions, inverse operations can be very helpful too!

$$\frac{4}{10} \qquad\qquad \frac{x}{15}$$

Cross-multiply $4 \times 15 = 10 \times x$

Combine terms $\dfrac{60}{10} = \dfrac{10x}{10}$

Use inverse operations

$x = 6$

Practice

Substitute and solve the following expressions for $b = 7$, $c = \dfrac{1}{2}$, and $f = 3$.

1. $7c + 2$

2. $f \times b$

3. $3 + f + c$

4. In which of these equations does $x = 6$?
   A. $3x = 2$
   B. $\dfrac{1}{2}x = 6$
   C. $5 + x = 11$
   D. $\dfrac{2}{3}x = 9$

Answers

1. $7 \times \dfrac{1}{2} + 2 = 3\dfrac{1}{2} + 2 = 5\dfrac{1}{2}$

2. $3 \times 7 = 21$

3. $3 + 3 + \dfrac{1}{2} = 6\dfrac{1}{2}$

4. **Only in C does $x = 6$.**

PARCC Question

Sean is 7 years older than Nick. If *s* represents Sean's age, write an expression for Nick's age. What is Nick's age if Sean is 17 years old?

Answer Explained

Nick is younger. So we subtract. $s - 7$. We substitute 17 for *s* and get $17 - 7$. Nick is currently 10 years old.

Inequalities

We have been working in the past several sections with equations. This means that the number sentence contains an equal sign. Sometimes we must work with quantities that may not be equal. This is called an **inequality**. When two quantities are not equal, we can indicate this with several different signs.

$\neq$	Not equal, not the same
$<$	Less than, fewer than, up to
$>$	Greater than, more than, exceeds
$\leq$	Less than OR equal to, up to and including, no more than, at most
$\geq$	Greater than OR equal to, no less than, at least

It is important to understand the difference between less than ($<$) and less than or equal to ($\leq$). This is a difference of whether the answer is included or excluded. If I say that a child must be taller than 48 inches to ride a roller coaster, a child who is EXACTLY 48 inches will not be allowed on. However, if I say that a child must be at least 48 inches, then he or she would be allowed on the ride.

We write inequalities the same exact way that we write equations. The only difference here is that the statement is NOT balanced. The vocabulary for the operations are the same, but instead of terms that mean equals we use various inequality terms.

The total of a number *x* and 7 is less than 12:

$$x + 7 < 12$$

Twice the quantity of *q* is greater than 19:

$$2q > 19$$

The direction of the inequality matters. Unlike with equations we can't just swap the two sides in our answer. Think about it this way: 4 > 3 is true but if we swap the two numbers 3 > 4 it is now false. We can still switch the order of our answer. We just have to remember to change the sign as well.

If we get an answer of 14 > x we can rewrite that as x < 14. This makes sense because if 14 is bigger than x, obviously x is smaller than 14.

Just as in equations, we use inverse operations to isolate the variable on one side of the inequality.

$$x + 7 < 12$$
$$\underline{-7 \quad -7}$$
$$x < 5$$

This says that x can be ANY value that is less than 5. Answers could be 3, 1.2, $\frac{1}{3}$, and others.

$$2q > 19$$
$$\underline{\div 2 \quad \div 2}$$
$$q > \frac{19}{2} \text{ or } 9\frac{1}{2} \text{ or } 9.5$$

This means that q can be ANY value that is greater than $9\frac{1}{2}$. Answers could be 10, $9\frac{2}{3}$, 127, and others.

In working with inequalities, we see that there are so many possible answers. We leave the solution as x < 5 or $q > 9\frac{1}{2}$. We also can graph these answers on a number line. We need to show that there are many possible answers. A line shows us not only whole number answers but all the fractional parts between.

Before we start graphing our inequalities it is much easier if we are sure that our answer is a variable first, such as z < 3. If we have the answer first, flip it and then graph. For example: 9 < m becomes m > 9. This will prevent confusion because now the arrow on the graph goes in the direction of the inequality symbol.

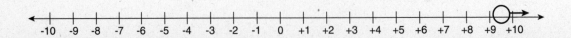

Notice that both are started by an open circle. An open circle indicates < or >.

Let's graph $x \geq -2$ and $x \leq 6$.

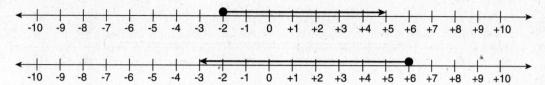

When we have a closed circle, the starting number is included in the number line, which is indicated by $\leq$ or $\geq$.

 **HELPFUL HINT**

Make sure that you are using the correct inequality symbol in the correct place; $x < 6$ is very different from $6 < x$. The first inequality is all answers less than 6; the second states that 6 is less than x, meaning the answer is all answers greater than 6, or $x > 6$.

Practice

Solve and graph on a number line.

1. $3 + x > 11$

2. $2y < 14$

3. $d - 2 \leq 6$

4. $\dfrac{m}{3} > 2$

Answers

1. $x > 8$

2. $y < 7$

3. $d \leq 8$

4. $m > 6$

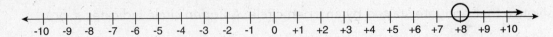

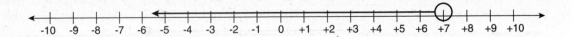

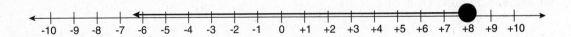

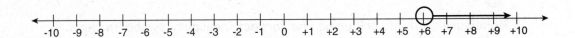

PARCC Question

Which of the following solution sets correctly identifies the inequality $2x < 7$?

○ **A.** {–2, –1, 0, 6, 8}

○ **B.** {–4, –1, 0, 1, 2}

○ **C.** {3, 4, 5, 6}

○ **D.** {7, 14, 21, 28}

Answer Explained

$x < 3.5$ A, C, D all have answers that are greater than 3.5 so only B is the correct answer.

Independent and Dependent Variables

Often an equation is made up of two variables in relation to one another. These two variables are called the dependent and independent variables. An independent variable is a variable that we can control. A dependent variable is the result of the independent variable. It "depends" on it. There are lots of examples of independent and dependent variables. The amount of data you use on your cell phone is independent; the cost of the bill is dependent. The amount of hours you work is independent; your total paycheck is dependent. In a combination of independent and dependent variables, there are numerous (and sometimes infinite) solutions.

Let's say that John is 6 years older than his sister Sarah.

$$\text{Sarah} + 6 = \text{John} \quad \text{or} \quad s + 6 = j$$

Depending on the value we give s, the value for j changes.

The best way to start organizing these variables it to put possible answers on a table. Let's use easy numbers:

Sarah (s)	$s + 6 = j$	John (j)
2	2 + 6 = 8	8
4	4 + 6 = 10	10
6	6 + 6 = 12	12
10	10 + 6 = 16	16

We can see that when Sarah was 2 years old, John was 8 years old; when she was 4 he was 10 and so forth. We can see there is a definite pattern.

Let's take these variables and instead of calling them s and j, let's change them to x and y. Remember it doesn't matter what letters we use as a variable.

With the variables x and y, we are able to create ordered pairs. (x, y)

We can turn the numbers on the table into the following ordered pairs:

$$(2, 8), (4, 10), (6, 12), \text{ and } (10, 16)$$

What do we do with ordered pairs? Plot them on a graph!

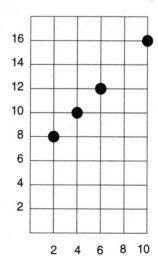

When we look at these dots, it is easy to see that they are arranged in a perfect line.

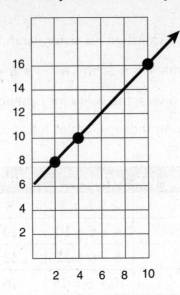

We can draw this line over an entire graph. This line now shows all of the possible combinations of their ages. For example, if Sarah is $7\frac{1}{2}$ John is $13\frac{1}{2}$ because it is on our line.

The line on the graph shows the complete answer (all answers) for $s + 6 = j$ ($x + 6 = y$).

 **HELPFUL HINT**

The most common questions about independent and dependent variables use common formulas such as Distance = Rate × Time.

Practice

Norah has 3 more marbles than her brother Tim. Write an equation of two variables that represents Tim's marbles. How many marbles does Norah have if Tim has 27? How many marbles does Tim have if Norah has 27?

Answers

$N - 3 = T$

Norah has 30 marbles.

Tim has 24 marbles.

PARCC Question

A factory makes cars at a regular rate. The rate can be written as $y = 18x$ where x is hours and y is number of cars made.

List the ordered pairs for 0, 1, 2, 3, and 4 hours.

(_, _), (_, _), (_, _), (_, _), (_, _)

Which variable is dependent? Which is independent?

Answer Explained

We find the y value by multiplying the x by 18.

(0, 0), (1, 18), (2, 36), (3, 54), (4, 72)

The x (hours) is independent, and y (cars) is dependent. The number of cars made depends on how many hours the factory is running.

Review Test

1. Rewrite as an exponent.

   $2 \times y \times 3 \times y \times f \times f \times y$

   $k \cdot k \cdot z \cdot z \cdot m$

   $4 \times 4 \times 4 \times 4$

   $6 \times 6 \times 6$

2. Expand the exponent; solve if possible.

   6^4

   9^3

   2^5

   h^6

3. Rewrite the following expressions so that the coefficient is 5 and the constant is 3.

   $2x + 2$

   $8 - 8y$

   $4z + 7$

4. Match the expressions/equations with their word form.

   3x + 2   3 more than 2 higher than a number x

   2 − 3x   3 groups of x less 2

   3 + 2 + x   2 more than 3 times a quantity x

   3x − 2   3 groups of x less than 2

5. Write the equation with b as the variable.

   Lucy has 8 packages of hamburger rolls for a barbeque. One of the packages is missing 4 buns.

6. Use the Distributive Property to write equivalent expressions.

   3x + 6

   6 (2r − 3)

   8 + 12y

7. Substitute and solve for $r = 1\frac{1}{2}$, s = 3, and t = 8.

   4r + 2s + 8t

   $\frac{1}{2}t - r$

   $s^3 + t^2$

8. Solve for x using inverse operations.

   4x − 5 = 27

   45 + 3x = 66

   4x + 2x = 54

9. Solve the proportions using inverse operations.

   $\frac{4}{6}$ $\frac{x}{24}$

   $\frac{3}{8}$ $\frac{15}{x}$

   $\frac{x}{8}$ $\frac{12}{32}$

10. Write and graph an inequality.

   For the temperature of a freezer that needs to be under 25 degrees.

   For a passing test grade if passing starts at 60.

   For a car with a top speed of 115 miles per hour.

11. Solve the inequalities.

    $3x + 3 > 9$

    $2x - 8 < 42$

    $8 + 4x > 20$

12. Solve for y when $x = 3$.

    $2 + 7x = y$

    $5x + 9 = y$

    $2y - 5 = x$

13. Complete the table and the graph for the following chart.

X	X + 4	Y

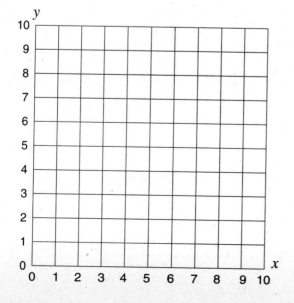

Answers

1. $2 \times 3 \times y^3 \times f^2 = 6y^3f^2$

    $k^2 \times z^2 \times m$

    4^4

    6^3

2. $6 \times 6 \times 6 \times 6 = 1,296$

    $9 \times 9 \times 9 = 729$

    $2 \times 2 \times 2 \times 2 \times 2 = 32$

    $h \times h \times h \times h \times h \times h$

3. $5x + 3$

    $3 - 5y$

    $5z + 3$

4. $3x + 2$ 2 more than 3 times a quantity x

    $2 - 3x$ 3 groups of x less than 2

    $3 + 2 + x$ 3 more than 2 higher than a number x

    $3x - 2$ 3 groups of x less 2

5. $8b - 4$

6. $3(x + 2)$

    $12r - 18$

    $4(2 + 3y)$

7. $4(1\frac{1}{2}) + 2(3) + 8(8) = 6 + 6 + 64 = 76$

    $\frac{1}{2}(8) - 1\frac{1}{2} = 4 - 1\frac{1}{2} = 2\frac{1}{2}$

    $s^3 + t^2 = 3^3 + 8^2 = 27 + 64 = 91$

8. $4x - 5 = 27$ $45 + 3x = 66$ $4x + 2x = 54$

    $4x = 27 + 5$ $3x = 66 - 45$ $6x = 54$

    $x = 32/4$ $x = 21/3$ $x = 54/6$

    $x = 8$ $x = 7$ $x = 9$

9. $\dfrac{4}{6}$ $\quad \dfrac{x}{24}$ $\qquad 4 \times 24 = 6x \qquad 96 = 6x \qquad 96/6 = x = 16$

$\dfrac{3}{8}$ $\quad \dfrac{15}{x}$ $\qquad 8 \times 15 = 3x \qquad 120 = 3x \qquad 120/3 = x = 40$

$\dfrac{x}{8}$ $\quad \dfrac{12}{32}$ $\qquad 8 \times 12 = 32x \qquad 96 = 32x \qquad 96/32 = x = 3$

10.

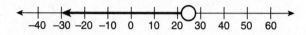

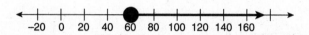

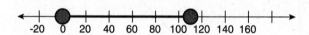

Notice that the third line has two stopping points; the car cannot go less than zero either.

11. $3x + 3 > 9$
$3x > 9 - 3$
$x > 6/3$
$x > 2$

$2x - 8 < 42$
$2x < 42 + 8$
$x < 50/2$
$x < 25$

$8 + 4x > 20$
$4x > 20 - 8$
$x > 12/4$
$x > 3$

12. Solve for y when $x = 3$

$2 + 7x = y$
$2 + 7(3) = y$
$2 + 21 = y$
$23 = y$

$5x + 9 = y$
$5(3) + 9 = y$
$15 + 9 = y$
$24 = y$

$2y - 5 = x$
$2y - 5 = 3$
$2y = 3 + 5$
$y = 8/2$
$y = 4$

13. Complete the table and the graph $x + 4$.

X	$X + 4$	y
0	$X + 4$	4
1	$X + 4$	5
2	$X + 4$	6
3	$X + 4$	7

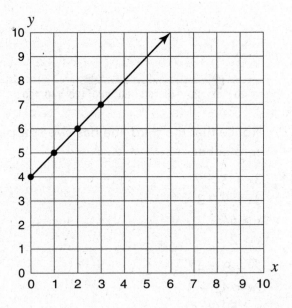

Geometry

Geometry is the study of shapes. In the sixth grade, it is about measuring these shapes, focusing on the area of various polygons, especially triangles. We are going to look at how other polygons are really just a combination of rectangles and triangles. We are going to use what we know about measurement to find the volume and the surface area of a prism.

Area of Triangles

If we start looking at right triangles (triangles with a right angle), we can easily see that they are simply one half of a rectangle.

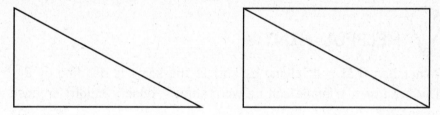

After we identify the rectangle, it is easy to find area. In a rectangle, we already know that area is length × width. In a triangle, we are going to call those two sides base and height. The base and height always meet at a right angle.

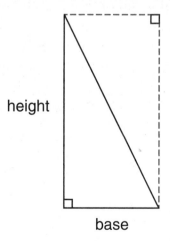

height

base

If we were finding the area of the rectangle, we would be multiplying the base by height, The triangle is half of the rectangle, so if we solve for the rectangle and divide by 2, we see that we have found the area of the triangle. When we divide the rectangle this way, it can be said that we **decompose** it. Decompose is the act of splitting something into smaller parts. In this case, we decompose a rectangle into two triangles. I can **compose**, or put together, two equal right triangles to make a rectangle.

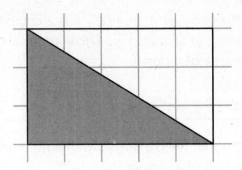

 **HELPFUL HINT**

Remember that multiplying by half is the same is dividing by 2. The area of a triangle can be written as $\frac{1}{2}$ base × height or base × height ÷ 2.

You may be worrying about what we do for all triangles that are not right triangles. It is easy! All triangles can be divided, or decomposed, into two right triangles. The right triangles can be turned into rectangles. If we take one side (the base) and draw an imaginary perpendicular line at a right angle (the height) from the highest point to the base, we have created two right triangles.

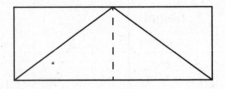

This allows us to use the same formula. The line you have drawn is now the height!

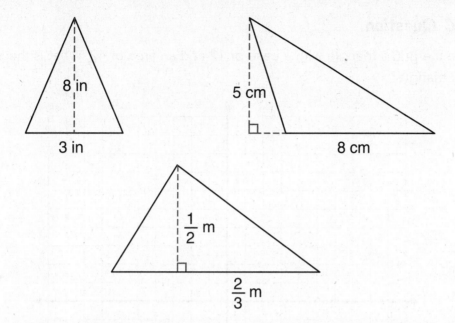

We can use the measurements in these triangles to find area. Don't forget that just like rectangles we write the label squared.

$$8 \times 3 \div 2 = 12 \, \text{in}^2 \qquad \frac{1}{2} \times 8 \times 5 = 20 \, \text{cm}^2 \qquad \frac{1}{2} \times \frac{1}{2} \times \frac{2}{3} = \frac{2}{12} = \frac{1}{6} \, \text{m}^2$$

Practice

1. What is the area of a triangle with a base of $6\frac{1}{2}$ and a height of $1\frac{1}{3}$?

2. What is the base of a triangle with a height of 5.7 and an area of 6.3?

3. What is the height of a triangle with a base of 6 and an area of 27?

Answers

1. $6\frac{1}{2} \times 1\frac{1}{3} \div 2 = 4\frac{1}{3}$ square units; we divide by 2 because we are finding area.

2. $6.3 \div 5.7 \times 2 = 2.21$ units; we multiplied by 2 because we are finding the base and need to do the opposite of finding area.

3. $27 \div 6 \times 2 = 9$ units; we multiplied by 2 because we are finding the height and need to do the opposite of finding area.

PARCC Question

Draw in the grid a triangle with a base of 12 and an area of 42. What is the height of your triangle?

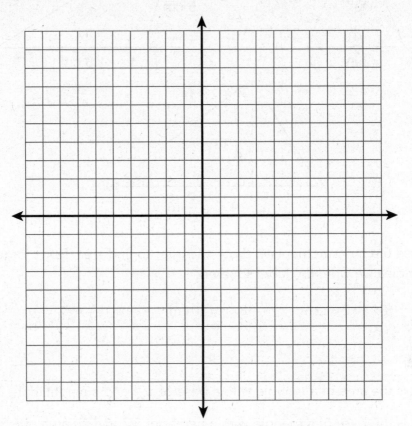

Answer Explained

Many varieties of drawings. Including

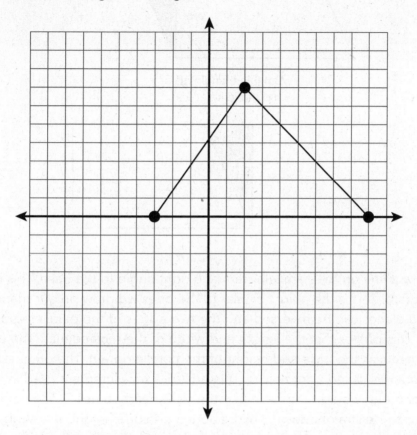

The height is 7. Any illustration that has a base of 12 and a height of 7 is accurate.

Area of Parallelograms, Trapezoids, and Other Polygons

Other than rectangles and triangles, there are many other polygons. First, we have the parallelogram. Parallelograms are figures with 2 sets of parallel lines. A rectangle is a special parallelogram that has 4 right angles. Parallelograms are actually made out of rectangles.

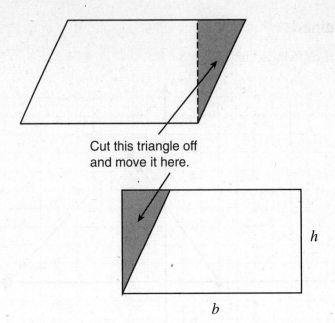

Cut this triangle off
and move it here.

h

b

We draw a line on the parallelogram at a right angle to the base. This cuts the parallelogram. This is the height similar to the ones we draw on our triangles. These two pieces can be reversed and the two edges of the parallelogram fit perfectly. This proves that the length and width of the rectangle are the exact same measurements as the base and height of the parallelogram. The other side of a parallelogram is called slant height. This is NOT a measurement used in area, but to find perimeter we add the base and the slant height times 2. The height we use in area has to be straight down. Think about a person's height; it is straight down to the floor.

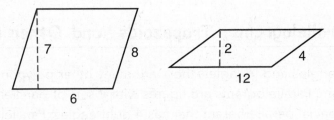

$$6 \times 7 = 42 \quad 12 \times 2 = 24$$

Another common quadrilateral is a trapezoid. A trapezoid is a figure with one set of parallel sides. These parallel sides are called bases (base 1 and base 2). Here we draw the height twice (once at each vertex of the smaller base).

By drawing these two height lines we have decomposed the shape into a rectangle and two right triangles.

We find the area of the rectangle and each triangle, and then we add the three pieces together. In this case the triangles are the same but they don't have to be equal.

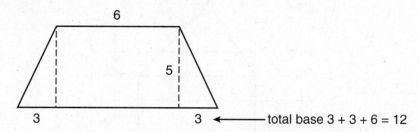

The rectangle is 6 × 5 = 30. The two triangles are identical with a base of 3 and a height of 5. 3 × 5 = 15 (we didn't divide in half because we have two of these triangles.) We add the pieces together and have an area of 45 square units.

💡 HELPFUL HINT

There is a formula for area of a trapezoid. It will **not** be provided to you because PARCC wants you to find the area using triangles and rectangles. The formula is $A = \frac{\text{Top Base} + \text{Bottom Base}}{2} \times \text{Height}$. Memorizing this can help you check your work.

Lots of other polygons can be decomposed into triangles and rectangles. By measuring each of these figures and finding their individual areas, they can be added together to find the area of the whole figure.

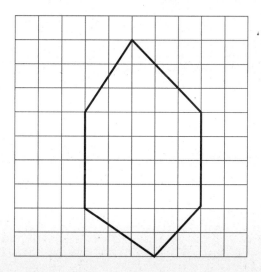

When breaking down figures, be sure that you are careful about their measurements. Be sure you are looking at the length of the sides carefully. Draw lines that make measurements that align with the grid, or are even units, whenever possible.

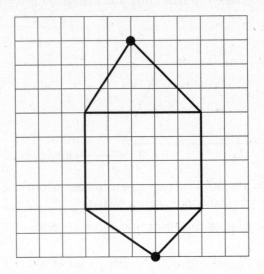

HELPFUL HINT

These polygons may be drawn on a coordinate plane with grid lines. Do not be tempted to count the squares for an accurate measurement of the area (it can be used to estimate).

Practice

Find the area of the following figures.

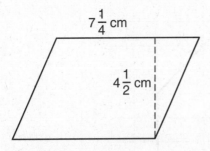

$7\frac{1}{4}$ cm

$4\frac{1}{2}$ cm

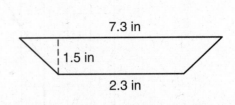

7.3 in

1.5 in

2.3 in

Answers

The area of parallelogram is $32\frac{5}{8}$ square centimeters.

The area of trapezoid is 7.2 square inches.

PARCC Question

The image shows the pattern on the tile floor in the lobby of a hotel. The tiles are each 1 square inch. The triangles are all identical and are green. What is the total area of green tile in square inches?

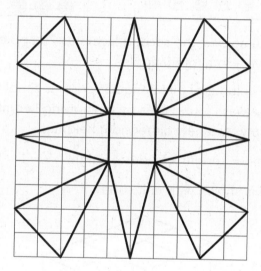

Answer Explained

There are 8 green triangles. Many of the triangles are hard to read but you can use the easier triangles to tell that each triangle has a base of 2 and a height of 4.

$\frac{1}{2}(2)\times4=4$ square units. There are 8 triangles for a total of 32 square inches.

Volume of Rectangular Prisms

We have done quite a bit with understanding area. Area is looking at the size of a **two-dimensional**, or flat figure. Now we need to look at three-dimensional figures, or figures that have physical shape. We are going to start by measuring the volume of a three-dimensional figure. Volume measures how many cubes can fit inside a figure. In sixth grade, we only need to look at **rectangular prisms**. A rectangular prism is a three-dimensional figure that is made up of rectangles. Tissue boxes, cereal boxes, and bricks are all examples of rectangular prisms.

Let's imagine we have an apple crate. The crate has 3 layers of apples. Each layer holds 20 apples, in 4 rows of 5.

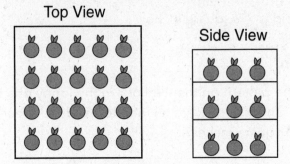

The first layer has 20 apples, the second layer has 20 apples, and the third layer has 20 apples. We have a total of 60 apples.

The formula for the volume of a rectangular prism is simply to multiply the three dimensions. We have always found area by multiplying Length × Width. But now we are going to multiply by a new dimension: height.

$$\text{Volume} = \text{Length} \times \text{Width} \times \text{Height} \quad V = L \times W \times H$$

Sometimes volume is shown with unit cubes.

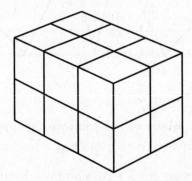

If this is the case you must count the height, the width, and the length and then multiply. Do not try counting all the cubes for the answer because some are on the "inside" and are hidden.

Other times you will be given a labeled picture.

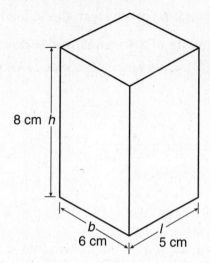

In this case you simply multiply the 3 dimensions: $5 \times 6 \times 8$.
5×6 is 30, and 30×8 is 240. Don't forget to label your answer. In area, we squared the units because we were counting squares, but with three dimensions we are looking at cubes so we cube our label! Our answer is 240 cubic centimeters or 240 cm³.

Cubes are a specific type of rectangular prism. Cubes are made of 6 identical squares. Dice, ice cubes, and sugar cubes are all examples of cubes. Since all sides are the same, the formula can be expressed two different ways: We can use the usual Length × Width × Height or instead use the formula Side³.

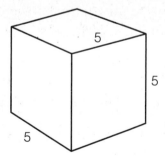

In this case, we can write $5 \times 5 \times 5$ or 5^3 or a total of 125 units³.

If we are given a key or are told that the cubes are of a fractional size, we must solve for the length, width, and height **before** we solve for area. For example, let's look at a rectangular prism with cubes that are $\frac{1}{2}$ inch long and measurements of 9 units long, 8 units high, and 5 units wide. We must first translate the measurements to $4\frac{1}{2}$ inches long, 4 units high, and $2\frac{1}{2}$ units wide before multiplying to find volume.

Practice

What is the volume of rectangular prisms with the following dimensions?

1. Length of 12 inches, height of 2.4 inches, and width of 1.7 inches

2. Length of 47 feet, height of 17 feet, and width of 12 feet

Answers

1. $12 \times 2.4 \times 1.7 = 48.96$ cubic inches

2. $47 \times 17 \times 12 = 9,588$ cubic feet

PARCC Question

If a rectangular prism has a volume of 144 cubic units, a length of 12, and a width of 3, what is the height?

Answer Explained

$$12 \times 3 \times H = 144 \quad 36 \times H = 144 \quad 144 \div 36 = 4 \text{ units}$$

Polygons on the Coordinate Plane

At the end of Chapter 3, we looked at plotting points on the coordinate plane. We also worked with measuring lines and finding points. In this chapter, we are going to expand on this to recognize that we use these lines to create polygons. We can use the coordinate plane to solve for perimeter and area of these figures.

First, we can work on reflecting points. When we reflect a point over an axis, we are keeping the point and changing the sign of one of the coordinates.

For example, if we have a point (3, 2) and reflect it over the *x* axis, we keep the *x* and change the sign on the *y* value and end up with (3, −2).

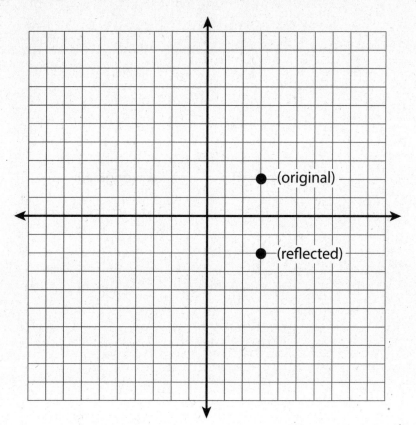

When we reflect it over the *y* axis, we keep the *y* and change the sign on the *x* value. For example, we go from (3, 2) to (−3, 2).

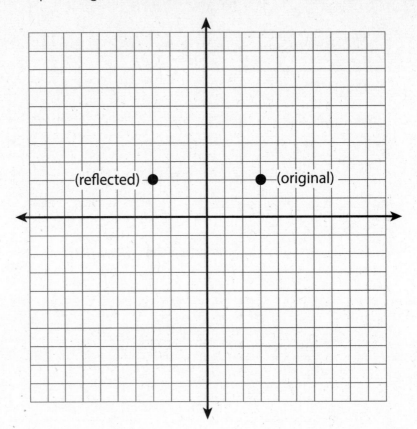

When we are told to reflect over the *x* axis and then the *y* axis (or the opposite, the *y* axis and then the *x*), we end up in the diagonal quadrant from the original point. This means changing *both* signs. (3, 2) becomes (–3, –2) and (–3, 7) becomes (3, –7).

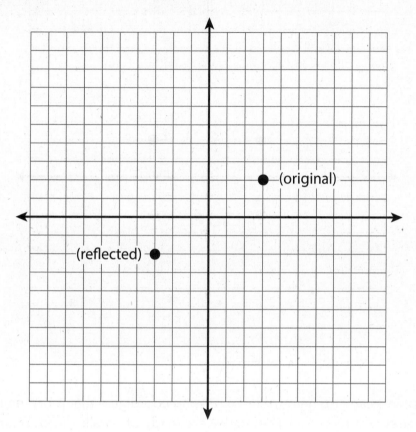

Together these four points create a rectangle that is centered on (0, 0) the origin.

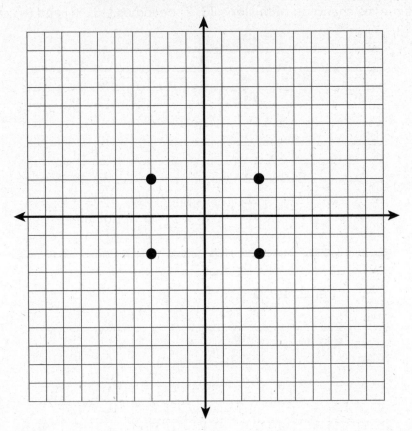

These points can also help us find a missing coordinate that represents a missing vertex of a rectangle. Rectangles are made of parallel lines. Therefore, if you have three coordinates, you can find the missing one, even without drawing them.

For example, we have (5, 6), (–3, –7), and (5, –7). What's the fourth?

If the lines are straight, they need to have a beginning and end point on the same line. The first and third point both are on $x = 5$, and the second point is $x = –3$; therefore, the fourth point must be $x = –3$. To find the y, do the same thing. Points 2 and 3 are both $y = –7$. The first and fourth must, therefore, be the same at $y = 6$.

Point 4 is $x = –3$ and $y = 6$, or the point (–3, 6).

We can prove this with a coordinate grid.

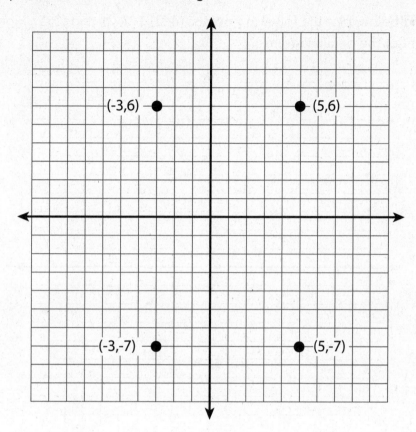

Using rules of reflection, we can find missing vertices of lots of different polygons. After we have all points of a polygon, we can calculate the area and perimeter of the polygon.

Practice

On the grid below, plot the following points: (4, 2), (–2, 2), and (0, 5). Plot the fourth point of the parallelogram.

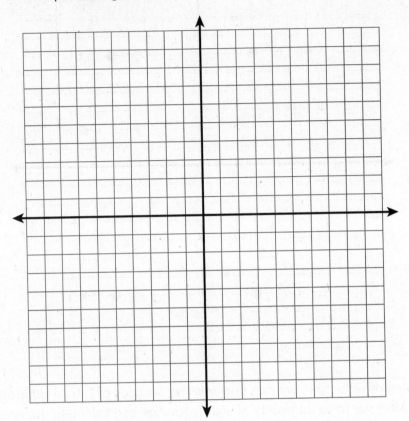

Answers

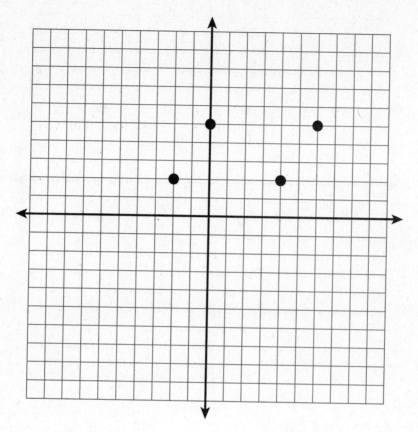

The fourth point is located at (6, 5).

PARCC Question

A right triangle is on a coordinate plane. The base of the triangle has two vertices located at (−2, 4) and (6, 4). It has a height of 6 units. Name two possible points for the third vertex.

Answer Explained

The third vertex has to make a perpendicular line from one of the other two vertices because it is a right triangle and therefore must have a right angle. Therefore, the point has to either be located at $x = −2$ or $x = 6$. It has to be 6 units from 4. If we go 6 units up, we are at $x = 10$. If we go 6 units down, we are at $x = −2$. There are 4 possible vertices (the problem asks for any 2): (−2, −2), (−2, 10), (6, 10), and (6, −2).

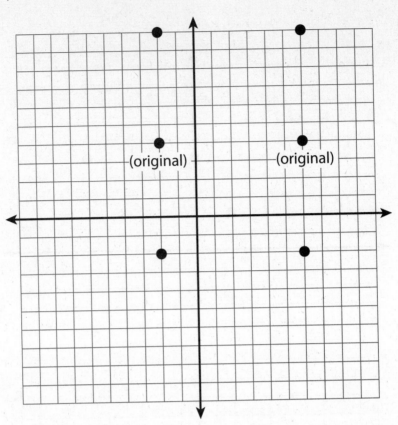

Surface Area and Nets

We still have one more geometric measurement that we need to review: surface area. **Surface area** is the calculation of the area of each face of a three-dimensional figure. Think about this as wrapping paper on a gift or the amount of paint on a box. We figure this out by solving for the area of each face and then adding all the faces together. The easiest way to organize it is to start by drawing a net. A **net** is a flat (two-dimensional) representation of a three-dimensional figure. Think of this as unfolding the shape.

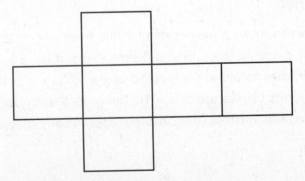

Prisms are made of two bases that are identical and rectangles that connect the bases. Since a rectangle has 4 sides, a **rectangle prism** has 2 bases and 4 rectangle faces. A triangle has 3 sides, so a **triangular prism** has 2 triangle bases and 3 rectangle faces. After we have the figures laid flat, we can start calculating the area of the faces.

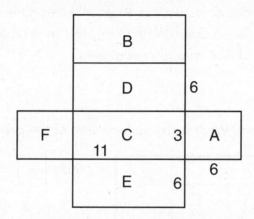

(A) 6 × 3 = 18   (B) 3 × 11 = 66   (C) 3 × 11 =33

(D) 6 × 11 = 33   (E) 6 × 11 = 66   (F) 6 × 3 = 18

Next we add all of the areas of the sides together.

$$A + B + C + D + E + F$$

18 + 66 + 33 + 33 + 66 + 18 = 234 square feet

Did you notice that *A* and *F* were the same? So were *C* and *D*, as well as *B* and *E*. Rectangles are made of equal opposite sides; because of this, opposite faces of a rectangular prism are also equal. There are 3 dimensions, or measurements, and each needs to be multiplied by each of the other two. There are 3 different combinations. We can use a shortcut when finding the surface area by finding the area of each of these three combinations and multiplying by 2.

 **HELPFUL HINT**

If the base is either a square or an equilateral triangle, each of the rectangles will be identical because the bases have equal sides. Don't do extra work. Solve the area of one rectangular face and multiply by the number of rectangles (4 for a square, and 3 for a triangle).

Practice

Find the surface area of a cube that has a side length of 12 centimeters.

Answer

A cube has all equal sides so it also has all equal faces. One face can be found by 12 × 12 = 144 square centimeters. We multiply this by 6 because there are 6 equal faces. We have a total of 864 square centimeters.

PARCC Question

Below is a net of a prism. What is the surface area of the prism?

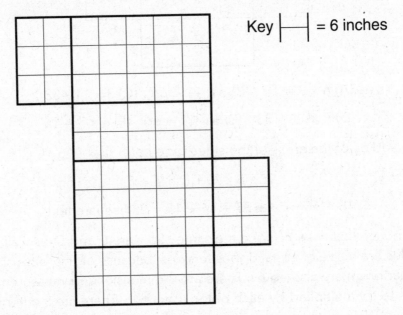

Key ⊢—⊣ = 6 inches

Answer Explained

The key says that each line is equal to 6 inches. (This makes each grid square 6 × 6 or 36 square inches.)

We count the three different size edges and get a height of 3 (3 × 6 = 18), a width of 2 (2 × 6 = 12), and a length of 5 (5 × 6 = 30). (It doesn't matter which dimension you call length, width, or height.)

We multiply Length × Height × 2, Height × Width × 2, Length × Width × 2.

$$30 \times 18 \times 2 = 1{,}080 \quad 18 \times 12 \times 2 = 432 \quad 30 \times 12 \times 2 = 720$$

We add it all together.

$$1{,}080 + 432 + 720 = 2{,}232 \text{ square inches}$$

Review Test

1. What is the area of a triangle with a base of 12 and a height of $3\frac{1}{2}$?

2. What is the base of a triangle with an area of $6\frac{1}{4}$ and a height of 2?

3. What is the height of a triangle with an area of $4\frac{2}{3}$ and a base of 7?

4. Find the area of the following triangle.

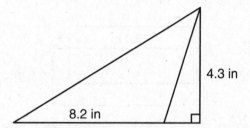

4.3 in

8.2 in

5. Draw a triangle with a base of 6 and an area of 4. Label its height.

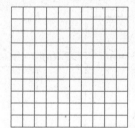

6. Find the area and perimeter for the parallelogram.

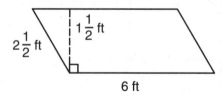

$2\frac{1}{2}$ ft

$1\frac{1}{2}$ ft

6 ft

7. Jackie has a parallelogram with a base of 5 and an area of $7\frac{1}{2}$. Find the height. Gerry has a parallelogram with the same area but a longer base. Is this possible? If it is possible, what could the base and height of his parallelogram be?

8. Find the area and perimeter for the trapezoid.

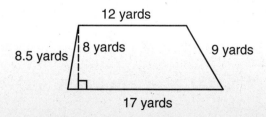

12 yards

8.5 yards

8 yards

9 yards

17 yards

9. Find the area of the trapezoid.

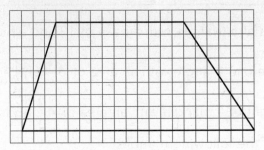

10. What is the volume of the figure below?

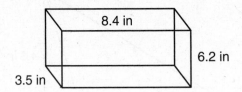

11. The net below folds into a rectangular prism. What is the volume of the prism?

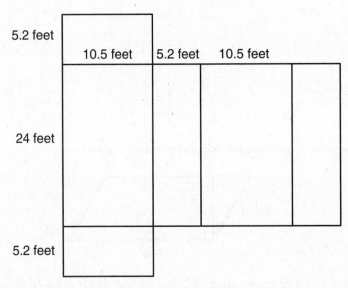

12. The volume of a box truck is 6,750 cubic feet. The width and height of the truck makes a 15 by 15 square. How long is the truck?

13. Find the surface area of the rectangular prism.

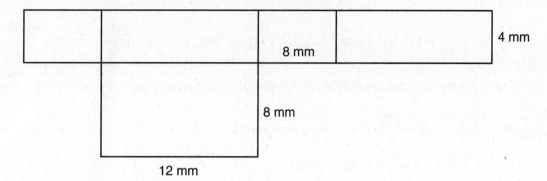

12 mm

14. Drew's Desserts needs to frost a rectangular sheet cake for a birthday party. The cake is 28 inches long, 15 inches wide, and 6 inches tall. A batch of frosting covers 300 square inches. How many batches of frosting do they need? (Remember they do not frost the bottom of the cake.)

15. Gary plotted a point at (−3, 2). His teacher said it was the incorrect answer and the correct answer was its reflection over the *y*-axis. What is the correct point?

Answers

1. 21

2. $6\frac{1}{4}$

3. $1\frac{1}{3}$

4. $b \times h \div 2 = 8.2 \times 4.3 \div 2 = 17.63$ square inches

5. There are many different drawings with a height of $1\frac{1}{3}$. One example:

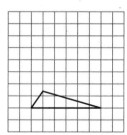

6. Area $= 6 \times 1\frac{1}{2} = 9$ square feet

   Perimeter $= 6 + 6 + 2\frac{1}{2} + 2\frac{1}{2} = 17$ feet

7. $7\frac{1}{2} \div 5 = b$

   $\frac{15}{2} \times \frac{1}{5} = \frac{15}{10} = 1\frac{1}{2}$ The base of Jackie's figure. Yes, you can have different

   measurements. One possibility is a base of 6 and a height of $1\frac{1}{4}$.

8. $A = \frac{12 + 17}{2} \times 8 = \frac{29}{2} \times 8 = 116$ square yards

   $P = 12 + 9 + 17 + 8.5 = 46.5$ yards

9. $\frac{11 + 20}{2} \times 9 = \frac{31}{2} \times 9 = 139\frac{1}{2}$ square units

10. $8.4 \times 3.5 \times 6.2 = 182.28$ cubic inches

11. $5.2 \times 10.5 \times 24 = 1{,}310.4$ cubic feet

12. $15 \times 15 = 225$

    $6{,}750 \div 225 = 30$ feet

13. $4 \times 12 + 4 \times 12 + 8 \times 4 + 8 \times 4 + 8 \times 12 + 8 \times 12$

    $48 + 48 + 32 + 32 + 96 + 96$

    352 square millimeters

14. $28 \times 15 + 15 \times 6 + 15 \times 6 + 28 \times 6 + 28 \times 6$

    $420 + 90 + 90 + 168 + 168$

    The cake needs 936 square inches frosted. $936 \div 300 = 3$ R 36

    You need a 4th batch of frosting to have enough.

15. Reflect over the y-axis means left to right. The y value stays the same and we have the reverse of the x value. (−3, 2) becomes the correct answer (3, 2).

Statistics and Probability

Statistics is the study of collecting, organizing, and drawing conclusions about **data**. It is used in a variety of ways including surveys, marketing, weather, and business. It allows us to find patterns and make predictions. We can compare information and make educated decisions. It can help companies determine how much product to stock, farmers decide the best fertilizer for their crops, and you decide if you need to pack a sweater on your vacation. In the study of statistics, we need to be able to pose statistical questions, collect data, summarize data with calculations and displays, answer questions, and draw conclusions about our summaries.

Statistical Questions

The first step in working with statistics is deciding if the questions are statistical in nature. Not all questions have a statistical answer. Questions that have a single fact-based result are NOT statistical. **Statistical questions** are ones that have **variability**, or different possible answers. For example, the number of inches in a foot is 12. There is no variety of answers. If we want to know how tall a sixth grader is, our answers will be numerous and varied. If we want to know what the current temperature is in the classroom, there is a definitive answer. If we want to know what the temperature is in all classrooms, we will get various readings, which makes this a statistical question. Statistical questions have two different types: numerical and categorical. **Numerical** questions have a number answer, including degrees and percent. **Categorical** questions have answers in words such as eye color, favorite flavor, and month.

Practice

Determine if the questions below are statistical. Mark as Y or N.

1. What is a comfortable temperature in a room?

2. What is the world record long jump?

3. What is the age when men get married?

4. How many grams of sugar are in a 6-ounce Hershey bar?

5. How many days are in August?

6. How many people like strawberry ice cream?

7. What is the typical number of students in a class?

Answers

1. **Y**
2. **N**
3. **Y**
4. **N**
5. **N**
6. **Y**
7. **Y**

PARCC Question

Part A

Your teacher has finished grading your test. Grades will post on Monday, but he said that he will answer statistical questions about the grades. At the first non-statistical question he is asked, he will stop answering any questions. Which of the following questions will you ask your teacher?

1. What was the average grade?

2. Did we do better on this test or the one for the last chapter?

3. How many students got an A?

4. How much better did our class do than all your past classes?

5. Who got the highest grade?

6. How many questions did we get wrong on the test?

7. What was the typical grade?

Part B

For any of the questions you would not ask your teacher, explain why they are not statistical.

Answer Explained

Part A

Your teacher would be willing to answer questions 2, 4, 6, and 7 because they are all statistical.

Part B

Your teacher would not answer question 1 because there is a single mathematical answer of the average. If you asked without the word average, it would be statistical.

Your teacher would not answer question 3 because there is a definitive number of As.

Your teacher would not answer question 5 because the highest grade is a fact.

Measures of Central Tendency

When looking to make observations about a group of numbers, the first step is often to put the data points in order from least to greatest. If a number appears multiple times, it should be included in the list multiple times.

 HELPFUL HINT

When putting the numbers in order, count the data points you started with and are on your list. Make sure you have the same number of data points in both.

If we have

$$3.7, 2.1, 2.5, 2.1, 3.2, 2.5, 2.5$$

Then we would order them as

$$2.1, 2.1, 2.5, 2.5, 2.5, 3.2, 3.7$$

When the numbers are in order, it is much easier to calculate for various measures of center. **Measures of center** help us draw conclusions about what is typical for a set of data. We are trying to determine what is true for the majority of the information. Understanding the center of data allows us to draw conclusions about a group as a whole. There are three main measures of central tendency.

One measure of center is the **median**. The median is the middle number. Median sounds like medium, another name for middle. If you list the data points in order, the data point in the center is your median. Think of this as lining up a choir from the tallest to the smallest singer: The person in the middle of the line has your median height. If there is an even number of data points (giving two numbers in the middle), we find the average of those two numbers.

For example,

Data: 2.7, 4.6, 3.8, 7.1, 5.6, 4.9

Data in order: 2.7, 3.8, 4.6, 4.9, 5.6, 7.1

The center is between 4.6 and 4.9. We find the average of these two numbers (4.6 + 4.9 = 9.5 ÷ 2 = 4.75); 4.75 is the median of this data set.

The second measure is mode. **Mode** is the number that appears most often. In looking at a series of data points, the mode is the data point that appears most frequently. There can be multiple modes if different data points are repeated the same number of times. You can think of this as the most common. If you ask a group of people how many pets they have, the answer you get the most would be the mode.

The mode is one measure of center that we can find with categorical data. If we ask a group of students their favorite band, we can't put them in order, but we can determine what answer is the most common, or the mode.

HELPFUL HINT

There will not always be a mode. If each data point is unique, no mode will exist.

For example,

0, 0, 1, 1, 2, 2, 2, 3, 3, 3, 4, 7

The mode(s) of this data is 2 and 3 because they both occur 3 times.

Mean, another name for average, finds another kind of middle. To find the mean, we start by adding all the data points. After we have a total, we divide by the number of data points that were collected. Averages are often used to find grades and to discuss typical costs.

For example,

4, 5, 6, 7, 6, 4, 5

We start by adding

$$4 + 5 + 6 + 7 + 6 + 4 + 5 = 37 \div 7 = 5.285\ldots$$

Practice

Find the mean, median, and mode for each set of numbers

1. Set 1: 4.7, 14.7, 2.4, 4.7, 3.8, 4.2

2. Set 2: 24, 4, 37, 56, 6, 56, 45

Answers

1. Set 1: Mean = 5.75; Median = 4.45; Mode = 4.7

2. Set 2: Mean = 32.57; Median = 37; Mode = 56

PARCC Question

The coach of a travel basketball team is trying to determine who gets the last spot on the team.

Paul has scored

11	9	10	11	8
10	12	9	11	10
11	10	11	10	9

Phillip has scored

2	18	7	12	6
6	18	14	9	5
11	14	7	20	3

What is the mean and median for each player?

Does mean or median give a better understanding of the player?

Who should the coach recruit if he wants the best scorer? Why?

Answer Explained

First, we have to be sure we are reading the chart correctly.

Paul has scored 8, 9, 9, 9, 10, 10, 10, 10, 10, 11, 11, 11, 11, 11, and 12.

Phillip has scored 2, 3, 5, 6, 6, 7, 7, 9, 11, 12, 14, 14, 18, 18, and 20.

Paul's

Mean: 8 + 9 + 9 + 9 + 10 + 10 + 10 + 10 + 10 + 11 + 11 + 11 + 11 + 11 + 12 = 152

152 ÷ 15 = 10.13

Median: The middle number is 10.

Phillip's

Mean: 2 + 3 + 5 + 6 + 6 + 7 + 7 + 9 + 11 + 12 + 14 + 14 + 18 + 18 + 20 = 152

152 ÷ 15 = 10.13

Median: The middle number is 9.

The mean is the same for both players, so the median is a better descriptor.

The coach should recruit Paul. His median is higher, and his scores are more consistent.

Measures of Variation

There are times when we are less concerned about the middle and more curious about how far and in what way the information is **spread**. What are my highest and lowest points? How does the information differ within the set? This is when we look at **measures of variation**.

We start with **range**. Range is simply looking at the highest data point and the lowest data point. Then we find the difference between the two.

High − Low = Range

 **HELPFUL HINT**

A large range tells us that the data is very spread out, and a narrow range tells us that the data is more clustered.

Interquartile range looks at the distance across the middle half of the information. To find the interquartile range, we first have to solve for each quartile. **Quartile** comes from the same root as the word *quarter*. It represents $\frac{1}{4}$ of the data. Earlier we discussed finding the median. The median is the middle of a group of numbers. This splits the data into two equal halves. If we want quarters, we will have to divide each of these pieces in half again, creating 4 equal parts.

Essentially, we are finding the median of a group of numbers, and then finding the median of each of these halves.

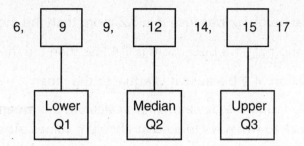

The interquartile range is best expressed as a box and whisker plot that we will build later in this chapter (see section on data displays).

The median of the first half is called the **lower quarter** or the Q1. It marks the division between the first and second quarter; it is halfway between the lowest point and the median.

The median of the second half is called the **upper quarter** or the Q3. It marks the division between the third and fourth quarters; it is halfway between the median and the highest point.

Remember just like when we find the median, we are looking at the middle. If there is an odd number of data points in the group, the middle number is easy to find. If there is an even number of points, we take the two numbers in the middle and find the average (mean) of the two numbers.

Once we have found the two quartiles we can find the interquartile range, "inter" meaning between, and "range" meaning difference.

To find the interquartile range we find the difference between the Q1 and Q3.

$$Q3 - Q1 = IQR$$

The interquartile range represents half of the data. The other half is split evenly between the first and fourth quarters.

Another way to study the variation or spread is the **deviation from the mean.** This sounds complicated, but we simply need to understand the vocabulary. We have already studied mean, and deviation means how far away. So deviation from the mean is found by calculating how far a given data point is from the average, or mean. It is a simple calculation of difference. The mean minus (subtract) the data point is the deviation from the mean. Deviation from the mean can be a positive number (if the data point is larger than the mean) or a negative number (if the data point is smaller).

If the mean of a data set is 23.1

21 has a deviation of −2.1 because it is 2.1 less than the mean.

24.3 has a deviation of 1.2 because it is 1.2 more than the mean.

21.7 has a deviation of −1.4 because it is 1.4 less than the mean.

23.1 has a deviation of 0 because it is equal to the mean.

Now that we have found the deviation, let's calculate the **mean absolute deviation**. Think back to when we looked at absolute values. *Absolute* means that it doesn't matter about the direction. Basically, we find the absolute value of the deviation. If the deviation is 6.2, the absolute deviation is 6.2. If the deviation is −3.8, the absolute deviation is 3.8. Finally we need to find the mean, or average, of these absolute deviations. To do this, you add all the absolute deviations together and divide by the number of data points.

First, we have a sample set of data. Let's try the height in inches of a group of students.

$$50, 51, 52, 52, 53, 55, 55, 58, 60$$

We start by finding the mean, just as we did in the section on measures of central tendency.

$$\frac{50+51+52+52+53+55+55+58+60}{9} = \frac{486}{9} = 54$$

54 is our mean, or our center. Now we need to find how far each number is away from 54, or the difference from 54. Remember that we need to find the absolute value of the deviation.

$$54 - 50 = |4| = 4$$
$$54 - 51 = |3| = 3$$
$$54 - 52 = |2| = 2$$
$$54 - 52 = |2| = 2$$
$$54 - 53 = |1| = 1$$
$$54 - 55 = |-1| = 1$$
$$54 - 55 = |-1| = 1$$
$$54 - 58 = |-4| = 4$$
$$54 - 60 = |-6| = 6$$

Now we find the mean of these absolute deviations.

$$\frac{4+3+2+2+1+1+1+4+6}{9}=\frac{24}{9}=2\frac{2}{3}$$

The mean absolute deviation is $2\frac{2}{3}$ or 2.66. Finding the mean absolute deviation is a lot of steps, but if you work carefully it is not that hard, and it gives excellent information about the spread of your data.

Practice

1. Find the quartiles and the interquartile range of the data

56, 48, 37, 72, 28, 33, 35, 42, 48

2. Find the mean absolute deviation of the data

29, 26, 34, 26, 33, 33, 18, 22

Answers

1. Median 42, Q1 = 34, Q3 = 52 IQR is 18.

2. The mean is 27.6. The MAD is 4.6.

PARCC Question

The tables tell the number of years of teaching experience the teachers in the math department at two different schools have.

Part A

Find the interquartile range of each.

Compare the results.

Part B

If one teacher from each building leaves, how would this change the data? Select one teacher from each and recalculate to support your results.

School A

11	2	3
19	16	9
4	8	7

School B

6	15	8
7	12	11
10	14	9

Answer Explained

Part A

For School A,

$$2, 3, 4, 7, 8, 9, 11, 16, 19$$

The median is 8 and the IQR is the difference between 3.5 and 13.5; 13.5 – 3.5 = 10.

For School B,

$$6, 7, 8, 9, 10, 11, 12, 14, 15$$

The median is 10, and the IQR is the difference between 7.5 and 13 (13 – 7.5 = 5.5)

School A has a wider range and a wider IQR. There is more spread in experience at School A.

Part B

Answers will vary. You should say that it depends on what teacher leaves how the distribution is shifted. The calculations should be accurate to the new data points, taking into account that there are now 8 teachers instead of 9.

Data Displays

Statistics is about analyzing data, and frequently the easiest way to do that is to create **visual models** of the information. Graphs, tables, and plots show a lot of information very quickly as well as the relationship between that information.

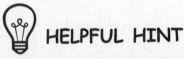 HELPFUL HINT

No matter what kind of display you are building, be sure to pay attention to all labels, titles, and scales. They matter!

Dot Plots

Dot plots are a way of displaying data. They show us the **frequency** that an answer occurs. They are a basic way of organizing information. Each data point is marked by an x or a mark in each category. It is important to have labels on the graph.

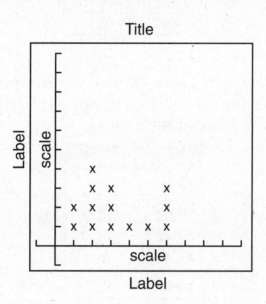

Histograms

You have made bar graphs for many years. A bar graph shows how many times a certain event occurs. You may have made bar graphs that show how many pets your classmates have or that illustrate election results. Each result gets its own bar. In a **histogram**, we start grouping answers. These groups are called **intervals**. Intervals divide a set of numbers into sections. Each interval represents all numbers between two defined points and one interval is adjacent to another, such as 1-5, 6-10, and 11-15. The histogram shows how many data points fall in a given category. Intervals must have the same amount of numbers. For example, you cannot have an interval of 0-5 (6 numbers) and one of 6-10 (5 numbers).

 **HELPFUL HINT**

Intervals cannot overlap. For example, with the intervals 0-5, and 5-10, 5 can only be in one interval.

Let's look at the following list of the ages of shoppers at a local store.

24	41	38	17	26
34	27	44	51	39
17	19	49	28	42
53	22	32	40	57
33	46	55	43	31

We can make our categories.

So we want to know how many ages are between 10 and 19, between 20 and 29, between 30 and 39, between 40 and 49, and between 50 and 59. We can first organize this information on a **frequency table**.

Age	Frequency
10-19	3
20-29	5
30-39	6
40-49	7
50-59	4

Now we are going to construct a histogram with this information. Be sure to include labels for both the *x* axis and *y* axis as well as a scale.

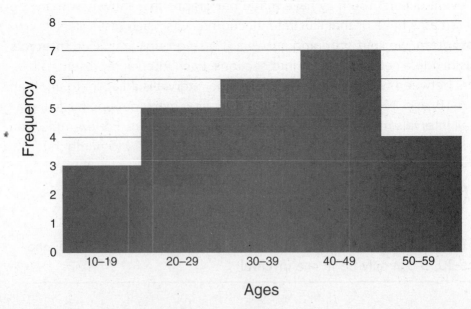

We can also answer questions about histograms.

Which category had the most shoppers? Which category had the fewest? What age is the center? How many shoppers were there?

HELPFUL HINT

When trying to categorize data points, we must be sure to add up frequencies in each category to get the original number of points. We don't want to forget anyone!

We learned earlier how to find the interquartile range using the upper and lower quartiles. A **box and whisker plot** is a way of displaying this information. First, we need a number line.

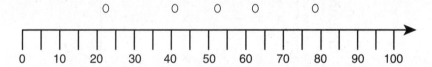

Then we plot the 5 numbers needed: the highest number (the maximum), the upper quartile (Q3), the median, the lower quartile (Q1), and the lowest number (the minimum).

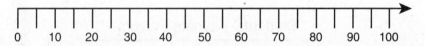

We draw a box between the upper and lower quartiles and split the box at the median. The box shows us the interquartile range. Next we draw a line or a whisker from the lowest number to the box and from the box to the highest number. The whiskers show us the full range.

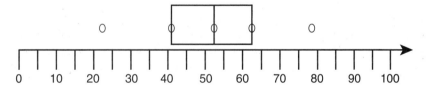

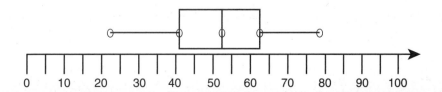

Practice

Draw the following data as a dot plot.

Julie asked her friends their favorite ice cream flavors. Her results are

Chocolate	10
Vanilla	6
Strawberry	3
Mint	2
Peanut Butter	3
Cookie Dough	9

Answers

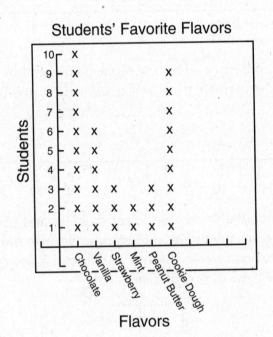

Students' Favorite Flavors

PARCC Question

A class is sprouting bean seeds for science class. The students each measure their sprouts in centimeters. They have measured their plants to be 6 cm, 5 cm, 8 cm, 9 cm, 6 cm, 4 cm, 5 cm, 8 cm, 6 cm, 7 cm, and 4 cm. Julie said that she built a box plot by starting at 4 and going to 9. She then drew a box from these two numbers. She split that box in half at $6\frac{1}{2}$ because that was the hallway point. Her box plot is below. How can you help Julie?

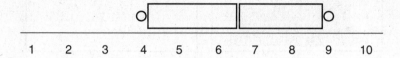

Answer Explained

The first step of using any data is to put the data points in order.

4 cm, 4 cm, 5 cm, 5 cm, 6 cm, 6 cm, 6 cm, 7 cm, 8 cm, 8 cm, 9 cm

Next, we find the median number which is the number in the middle of the group. The middle number of this data group is 6 cm. Now, we need to find the first quarter which is the middle of the first half or 5 cm. Then, we find the middle of the third quarter, which is the middle of the second half or 8 cm. We plot the 5 points and draw our box plot.

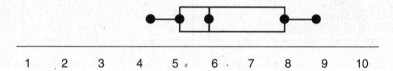

Describing Statistics

Now that we have practiced finding measures of center and measures of variability and have studied different types of graphs, there is one last step to understanding statistics. We now need to be able to describe the information. Many of these questions will be about calculating center and variability, but there will be other questions asking about the variability, and center as well.

We can use different displays and look for shape. **Shape** can be described in several ways. It can be described as centered. This is when the shape of the graph is mostly balanced. If it is unbalanced, it can be described as skewed. **Skew** is a shift to either the high end or the low end, showing that the information is unbalanced.

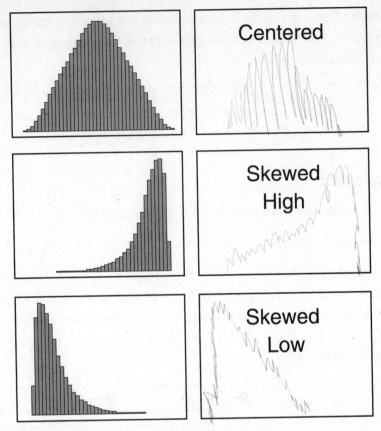

We can also describe statistics by mentioning any gaps, or points with no frequency between groups of high frequency.

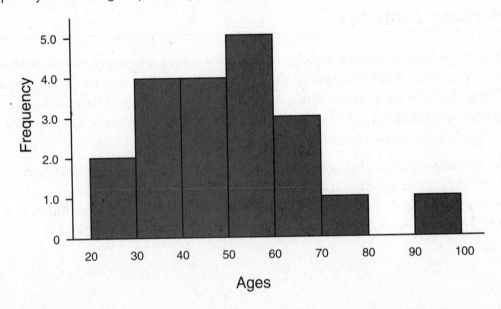

There are times where certain data points are separated from the majority of the data points. These **outliers** or unusual data points are still part of the measures of center and spread but it is important to note their distance from the common points. In this figure, 3 is the outlier.

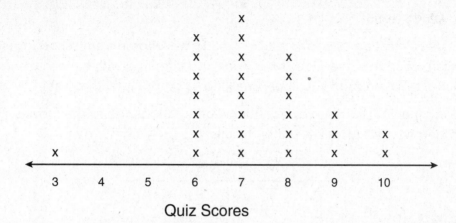

Quiz Scores

We can also discuss the number of data points that were collected overall. In comparing two different sets of information, it is important to note how many points are included. If one has a much larger sample than the other, comparisons can be more difficult.

Practice

Draw a picture of a data display for each of the following:

1. Outlier

2. Gap

3. High skew

4. Low skew

5. Balanced

Answers

Pictures will vary, but the key details are

1. One or two data points set apart from the rest

2. A set of data with a hole surrounded by a solid amount of data from each side

3. A set of data that has far more high numbers than low numbers

4. A set of data with far more low numbers than high numbers

5. A set of data with the highest data in the middle and equal amounts of low and high numbers

PARCC Questions

1. You are calculating how long it takes you to walk home from school. You measure your time every day for a week and come up with a mean of 17.2 minutes. How long did you spend walking to school total that week?

2. Two companies did a survey about how often people go to the movies. The following two dot plots show their results.

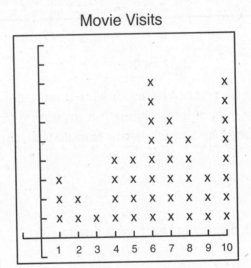

Movie Visits

Visits Per Year

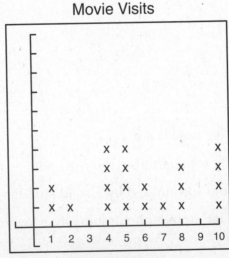

Movie Visits

Visits Per Year

Describe them individually using words such as mean, median, center, skew, and spread. Compare and contrast the two sets of results.

Answers Explained

1. Recognizing that we found mean by solving Mean = Total ÷ 5, we should be able to use total = Mean × 5 will help us find the full week. 17.2 × 5 = 86 minutes, or 1 hour and 26 minutes.

2. Descriptions will vary. Important information would include calculating the mean and median for each dot plot. They should note that the first study is slightly skewed to higher numbers. The second study is more balanced on center and clustered except for 2 zeros as outliers. The first study has more data points (more people surveyed).

Review Test

1. Write a numerical statistical question about soccer.

2. A magazine has 12 pages with the following numbers of words: 271, 354, 287, 314, 333, 326, 285, 327, 316, 301, 298, and 296. What is the mean number of words per page?

3. The age of 8 different visitors to a museum is 3, 13, 2, 34, 11, 26, 47, and 17. Find the median.

4. The cost (in dollars) of 9 different T- shirts are 18, 21, 11, 21, 15, 19, 17, 21, and 17. What is the mode price?

5. The month of April had a wide variety of temperatures. The recorded high was 82°, and there was a range in temperatures of 37 degrees. What was the recorded low temperature for the month?

6. Does the number of data points change the calculation of the range?

7. Denise is studying statistics and has to calculate the mean, median, and mode for a project. She keeps getting an answer for the median that is not on the list of values. Is this possible? Is it possible for the mode to not be on this list?

8. Find the mean, median, mode, and range for the time in seconds it took students to assemble the puzzle. 18, 23, 36, 24, 18, 19, 22, 24, 23, 21, 24, 18, 16, 24.

9. Find the median, Q1, Q3, and interquartile range for the miles traveled to school.

Miles Traveled to School			
5	24	5	21
17	16	11	9
13	18	23	22

10. The median score on a test was 83%. DJ had a 94% and Ashley had a 79%. Who had a greater absolute deviation from the mean?

11. Find the mean and mean absolute deviation for the number of vacation days taken by employees during the first and second quarter. Compare the mean absolute deviation of the two quarters.

Vacation Days Used First Quarter			
10	8	3	5
4	8	0	10

Vacation Days Used Second Quarter			
1	6	7	8
3	1	2	12

12. Construct a dot plot for the high school graduation dates of college applicants.

Graduation Year			
2012	2014	2011	2014
2011	2015	2010	2015
2014	2013	2014	2011

13. Create a histogram showing how many points the students improved between their pre-test and post-test. 8, 19, 25, 20, 9, 16, 12, 26, 7, 17, 33, 12, 4, 14.

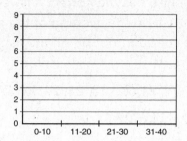

14. Answer the following questions about the box and whisker plot below.

What is the median?

What is the Q1?

What is the IQR?

What is the range?

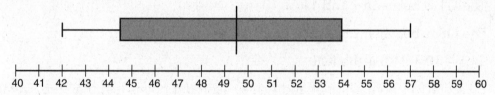

15. Draw a histogram with a **gap** and a **low skew**.

Answers

1. Lots of different answers. Questions should start with: How many? How long? How often? How old?

2. Added all together you get a total of 3,708 words or an average of 309 words per page.

3. First put the numbers in order: 2, 3, 11, 13, 17, 26, 34, 47.

   13 and 17 are the middle numbers: 13 + 17 is 30 ÷ 2 is 15. The median is 15.

4. 21 is the mode, or most common.

5. 82 – ? = 37 82 – 37 = 45. The low temperature was 44°.

6. No, as long as the highest and the lowest number don't change it doesn't matter if you have 4 or 40 data points.

7. Yes, it is possible that Denise has the correct answer. If there is an even set of data points, the median will be the average of the two middle numbers. No, it is not possible for the mode to not be on the list. Not only is it on the list but it is on the list the most!

8. In order: 16, 18, 18, 18, 19, 21, 22, 23, 23, 24, 24, 24, 24, 36

   The mean is 22.14. 310 ÷ 14 = 22.14 ...

   The median is between 22 and 23 or 22.5.

   The mode is 24. It appears 4 times.

   Range 36 – 16 = 20

9. 5, 5, 9, 11, 13, 16, 17, 18, 21, 22, 23, 24

   The median is between 16 and 17 or 16.5

   The Q1 is between 9 and 11 or 10

   The Q3 is between 21 and 22 or 21.5

   21.5 – 10 = 11.5 is the IQR

10. DJ scored 11 points higher and Ashley scored 4 points less than the average. Ashley is closer to the average, so DJ had a greater absolute deviation from the mean.

11. First Quarter. The mean is 6. The absolute deviations are 4, 2, 2, 2, 3, 6, 1, 4. The mean absolute deviation is 3.

    Second Quarter. The mean is 5. The absolute deviations are 4, 1, 2, 3, 2, 4, 3, 7. The mean absolute deviation is 3.25. The second quarter had a larger deviation.

12.

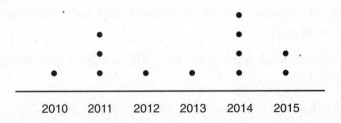

13.

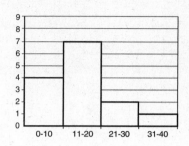

14.

What is the median? 49.5

What is the Q1? 44.5

What is the IQR? 54 – 44.5 = 9.5

What is the range? 57 – 42 = 15

15. One possible graph.

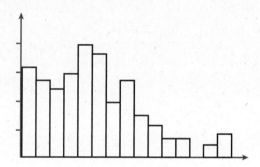

This is a sample of the test you will take on the PARCC. This sample is made up of all three types of questions: Type 1, Type 2, and Type 3. These questions will be assigned various point values depending on their difficulty and how many separate components you will need to answer. Some questions may or may not have a calculator. Some questions will have tools available to help you complete the task. Please refer to the information in Chapter 1 if you need more clarification.

> **IMPORTANT NOTE:** Barron's has made every effort to create sample tests that accurately reflect the PARCC Assessment. However, the tests are constantly changing. The following two tests differ in length and content, but each will still provide a strong framework for sixth-grade students preparing for the assessment. Be sure to consult *http://www.parcconline.org* for all the latest testing information.

Practice Test 1

1. Which of the following ratios of sugar to flour are equivalent to 2:3 in a cookie recipe? Check *all* that apply.
   - ☐ A. 2.6 cups of sugar: 3.6 cups of flour
   - ☐ B. 3.0 cups of sugar: 4.5 cups of flour
   - ☐ C. 212 cups of sugar: 318 cups of flour
   - ☐ D. 24 cups of sugar: 16 cups of flour
   - ☐ E. 12x cups of sugar: 18x cups of flour

2. Barbara's Bakery sells their mammoth muffins in several different size packages. You can buy a single muffin for $1.75; 6-packs are sold for $10.00; A dozen is $21.00; and a basket of 24 is $41.00. Find the unit price per muffin. Which is the best value?

3. Cheryl was visiting a friend for movie night, and they decided to organize her DVD library by genre.

Genre	Number of Movies
Horror	25
Romance	24
Comedy	36
Sci-Fi	20
Action	18
Animated	15

Part A

Complete the blanks to fit the relationship

The ratio of _____ movies to _____ movies is 3:4.

The ratio of _____ movies to _____ movies is 2:3.

The ratio of _____ movies to _____ movies is 3:5.

Part B

My friend only wants to watch a romance or a comedy. What is the ratio of the preferred movies to all movies?

Part C

Pick your three favorite movie genres and list them. What is the ratio of total movies to your favorite movies? What percent of my friend's movies are in your favorite three genres?

4. Kelly, her sister, and 4 brothers entered a snowman building contest. They were the second place team and earned $\frac{1}{4}$ of the prize money. What fraction of the prize money will each of them receive?

5. A fancy hotel was built into the side of a mountain. Many of the floors are built below ground level. They named these floors in negative numbers.
- The lobby is Level 0.
- The restaurant is located 2 floors above the guest floors.
- The maid service desk is located one floor above the pool.
- Guest rooms start 3 floors above and continue to 8 floors above the lobby.
- The Hotel gift shop is located one floor below the restaurant.
- The pool is located 5 floors below the guest room floors.

Which locations are listed as negative numbers? Pick the single best answer.

- ○ A. The gift shop and the pool
- ○ B. The pool and the lobby
- ○ C. The maid's desk and the pool
- ○ D. The guest rooms and the gift shop
- ○ E. The lobby, the pool, and the maid's desk
- ○ F. The pool only
- ○ G. None of these

6. Dwayne says that the negative of $-\frac{1}{4}$ is 4 because they are on opposite sides of the zero when they are on a number line. Steve disagrees. Which boy is correct? How would you fix the logic of the mistaken boy? Draw and label a number line to support your argument.

7. Carol decided to plot her neighborhood on a 4 quadrant coordinate plane. The neighborhood is split into north and south of Main Street, which she made the x axis, and into east and west of Broadway, which is her y axis. The origin of her plane is the intersection of these two roads. Each grid square equals one block.

Part A

Build and label the coordinate grid using these clues.

The library is at (7, 3).

The school is at (−9, 8).

The firehouse is at (7, −4).

The park is at (−8, −2).

The grocery store is at (−5, 6).

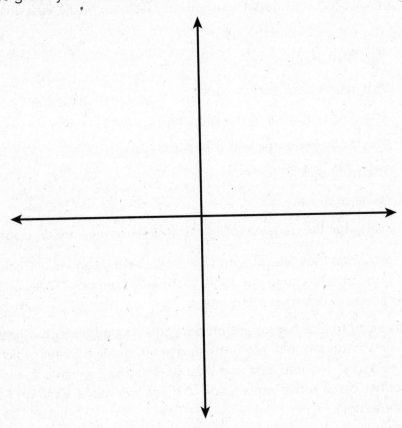

Part B

Some of the town did not fit on her original drawing, but she still wants to understand their relative distance.

The movie theater is located at (7, 17). How far is this from the firehouse?

The hospital is located at (−3, −2). How far is that from the park?

If the city wants to build a new post office exactly 6 blocks north of the school, where should they put it?

Part C

Draw the most direct path that you could take (no diagonals) between the school and the firehouse. What is the distance between the two?

8. Match the correct fractions on the number line. Put the following numbers in their correct place on the number line. Not all answers will be used.

$$1\frac{1}{5}, \ 1\frac{1}{2}, \ -1\frac{1}{5}, \ -1\frac{3}{6}, \ 1\frac{4}{5}, \ -1\frac{2}{3}$$

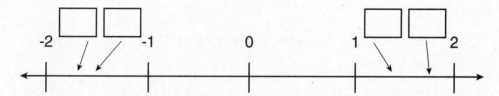

9. Which of the following are expressions equivalent to 6×3^4?

○ A. $6 \times 3 \times 4$

○ B. $6(3 \times 3 \times 3 \times 3)$

○ C. $3^4 \times 3^4 \times 3^4 \times 3^4 \times 3^4 \times 3^4$

○ D. $(6 \times 4) + (6 \times 4) + (6 \times 4)$

○ E. 6×12

○ F. $(6 \times 3) + (6 \times 3) + (6 \times 3) + (6 \times 3)$

10. **Part A**

Write and simplify an expression for the perimeter of a rectangle that has a length of $3c$ and a width of 2.

Part B

Write and simplify an expression for the area of a rectangle that has the same dimensions.

Part C

Find the perimeter and area of this rectangle if c is equal to 0.35 inches.

11. Leo, Emma, and Dave are going to the movies. Their cost can be described as three times the cost of a ticket and popcorn, or $3(t + r)$. Write this expression in another way.

12. **Part A**

Solve the following with correct order of operations. Show all your steps.

$$3 + 5^3 \times 6$$

Part B

Sean got the wrong answer, but all of his arithmetic was correct. Show a possible way of solving where Sean DIDN'T follow order of operations. What answer could he have gotten?

13. **Part A**

I have a cube. If the length of the cube is s, what is the width and height? How do you know? Draw a cube and label the dimensions.

Part B

Surface area is the total amount of area covering the outside of a three-dimensional figure.

Write an expression that will allow you to solve for the area of one face.

Use this expression to write and simplify an expression that expresses the entire surface area of the cube.

Part C

If Judy has a cube where s is equal to 3 cm, and Ben has a cube where s is equal to 4 cm, how much more surface area does Ben's cube have?

14. Decide if each statement is true, false, or not enough information when $x = 12$.

  ○ A. $6x > 72$

  ○ B. $9x > 99$

  ○ C. $30 - x + 5 = 13$

  ○ D. $3x - y = 17$

  ○ E. $6 + 2x = 30$

  ○ F. $5x + 2x = 84$

15. **Part A**

The cost to have a party at the local bowling alley is $250 for the space and $8.75 per participant ($p$). You also need to get shoes (s) at a cost of $2.15 for anyone who needs them. Write a two variable equation for the total cost of the party.

Party =

Part B

What is the cost of the party if you have 25 guests and only 8 of them brought their own shoes?

16. Victoria is saving to buy a new cell phone. She gets an allowance of $16.50 per week. She is hoping to save $198. She doesn't know how many weeks (*w*) this will take. Her brother Joe gives some possible equations to figure this out.

   ○ A. $16.50 + w = 198$

   ○ B. $w - 16.50 = 198$

   ○ C. $16.50 \times w = 198$

   ○ D. $w \div 16.50 = 198$

Part A

Which is the correct equation? Solve for *w*. Show your work.

Part B

Solve the other 3 equations. What does *w* equal? Explain if these answers make sense. How does that help you solve this problem?

Part C

Write an equation for if she gets an increase in her allowance to $18. What is the new value of *w*?

17.

Part A

The formula Mass × Acceleration = Force is a very common equation in physics. I have an object that has a mass of 6 lb. Use acceleration as your *x* value and force as your *y* value.

List ordered pairs (x, y) for acceleration of 1, 2, and 3.

(___, ___) (___, ___) (___, ___)

Part B

Use the coordinate points to create a graph illustrating the formula and its relationship on this coordinate grid.

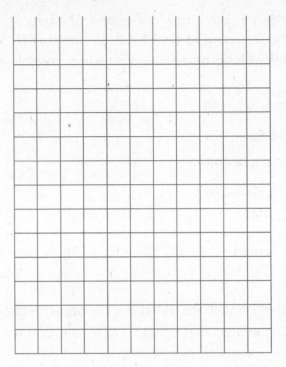

Part C

Explain what the independent and dependent variables stand for in this formula.

Answers

1. **B**, **C**, and **E** are equivalent answers. Choice B is increased by 1.5 cups; choice C is increased by 106 cups; and choice E is increased by 6x cups. Each of these represents a factor of 1.5. Choice A is often selected because students look only at the whole numbers. The numbers in option D are equivalent, but they are in the wrong order. 16:24 is equivalent but 24:16 is not. As long as the ratio of 2:3 is met, you can measure using anything! 2 shoe-fulls of sugar and 3 shoe-fulls of flour will give us the same dough!

2. To find the best value, you need to find the cost per muffin for each deal. We divide each cost by how many muffins it contains. The 6 pack is $1.66 per muffin; the dozen is $1.75 per muffin, and the 24 pack is $1.70 per muffin. The 6 pack is the best deal!

3. **Part A**

   Action to Romance is 3:4—18:24 (reduced by a factor of 6).

   Animated to Sci-Fi is 3:4—15:20 (reduced by a factor of 5).

   Romance to Comedy is 2:3—24:36 (reduced by a factor of 12).

   Animated to Horror is 3:5—15:25 (reduced by a factor of 5).

   Part B

   Romance and comedy together are 24 + 36 = 60.

   Total movies is 138.

   The ratio of preferred to total is 60:138, both divisible by 6 (10:23).

   Part C

   Answers vary. Be sure that you are adding correctly and that you reduce any unsimplified ratios. To find percent, divide your favorite total by 138 (the total number of movies). The answer will be a decimal, so move the decimal 2 points to the right and you have the answer as a percent!

4. 1 (Kelly) +1 (sister) +4 (brothers) = 6 kids

   $$\frac{1}{4} \div 6 = \frac{1}{4} \times \frac{1}{6}$$ (remember Keep, Change, Flip)

   They each get $\frac{1}{24}$ of the prize money.

5. C. The maid's desk and the pool

   The lobby is 0. The guest rooms are three floors above so they are floors 3-8. The restaurant is two floors above the guest rooms so it is on floor 10. The gift shop is one floor below the restaurant and, therefore, on floor 9. The pool is 5 floors below the guest rooms, so that is −2 and the maid's desk is one floor above the pool, which is −1.

   The answer of the maid's desk −1 and pool −2 is the best answer because they are both located on basement levels.

6. Steve is correct. The negative of $-\frac{1}{4}$ is not 4. If we draw a number line, we see that $-\frac{1}{4}$ is between 0 and −1. If we put 4 on the number line, it is on the other side of the zero, but it is MUCH farther away. The numbers have to be on two different sides of the zero but also at an equal distance. His answer should be $\frac{1}{4}$.

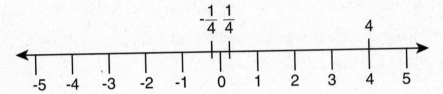

7. **Part A**

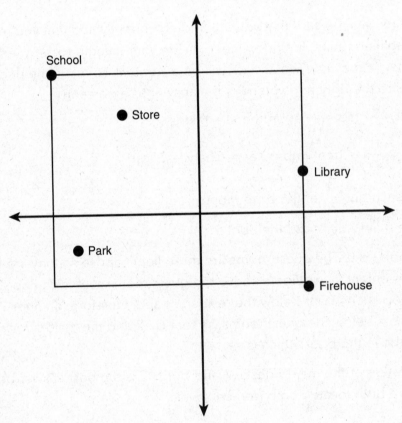

Part B

|−4| + |17| = 21 (different sign so we add)

|−8| − |−3| = 5 (same sign so we subtract)

If it is north, it will keep the same x coordinate of −9. 6 blocks north of 8 is 14. They will put it at (−9,14).

Part C

The student can either draw right from −9 to 7 (which is 16 blocks) and then down from 8 to −4 (an additional 12 blocks), a total of 28 blocks, or the student can go down first from 8 to −4 (12 blocks) and then go right from −9 to 7 (which is 16 blocks.) The total is still 28 blocks.

8. The following number line contains the answers. Notice that only 4 boxes are filled in; not all answers were used.

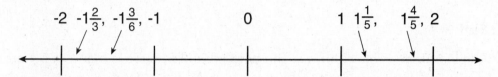

9. Only **B** is a correct answer. A and E are wrong because it says that 3^4 is 3×4 not $3 \times 3 \times 3 \times 3$. C, D, and F are wrong because you cannot distribute that way.

10. **Part A**

   $3c + 3c + 2 + 2$ 2 lengths and 2 widths

   $6c + 4$ combine your like terms

   $2(3c + 2)$ factor out the common 2

   Part B

   $3c \times 2$ (length times width)

   $3 \times c \times 2 = 3 \times 2 \times c$ Commutative Property lets us move the integers.

   $6 \times c = 6c$ combine like terms

   Part C

   Perimeter $= 2(3 \times 0.35 + 2) = 2(1.05 + 2) = 2(3.05) = 6.1$ units

   Area $= 6(0.35) = 2.1$ square units

11. Acceptable answers include distributed $3t + 3r$ or expanded $(t + r) + (t + r) + (t + r)$.

12. **Part A**

   $3 + 5^3 \times 6$

   $3 + 125 \times 6$ solve the exponent or $(5 \times 5 \times 5)$

   $3 + 750$ solve the multiplication

   753 finish with the addition

Part B

There are several ways that Sean could have set up his problem incorrectly. One way (and probably the most common) is to work left to right:

$3 + 5^3 \times 6$

$8^3 \times 6$

512×6

$3,072$

13. **Part A**

The length is s. The width is s. The height is s. The cube is made of squares. A square is a shape defined by having equal sides. So if one edge of a cube is s, the others are also all s.

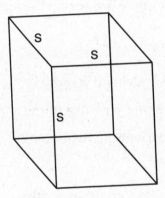

Part B

The area of a square is $s \times s$. There are six faces to add together, which is best written as the expression $6(s \times s)$.

The final step to the answer is recognizing that $s \times s$ is really s^2, making the simplest form of the expression $6s^2$.

Part C

We use the expression $6s^2$.

Judy $s = 3$ cm   Ben $s = 4$ cm

6×3^2 6×4^2

6×9 6×16

54 cm^2 96 cm^2

$96 - 54 = 42$

Ben's cube has a surface area that is 42 cm^2 larger than Judy's.

14.

    (A) $6x = 6(12) = 72$. It is not greater. This is false.

    (B) $9x = 9(12) = 108 > 99$. This is true.

    (C) $30 - x + 5 = 30 - 12 + 5 = 18 + 5 = 23$, which is not 13. This is false.

    (D) We cannot answer this, because we have no idea what is the value of y. Not enough information.

    (E) $6 + 2x = 6 + 2(12) = 6 + 24 = 30$. This is true.

    (F) $5x + 2x = 5(12) + 2(12) = 60 + 24 = 84$. This is true.

15. **Part A**

The expression would be

$$\text{Party} = 250 + 8.75(\text{participants or } p) + 2.15(\text{shoes or } s)$$

Part B

$$\text{Party} = 250 + (8.75 \times 25) + (2.15 \times 17) \quad \text{8 already own shoes so } 25 - 8 = 17.$$

$$\text{Party} = 250 + 218.75 + 36.55$$

$$\text{Party} = \$505.30$$

16. **Part A**

C is the correct equation.

We divide (opposite of multiplication) both sides by 16.50.

$$w = 198 \div 16.50 = 12$$

It will take her 12 weeks.

Part B

$$w = 198 - 16.50 = 181.5$$

$$w = 198 + 16.50 = 214.5$$

$$w = 198 \times 16.50 = 3{,}267$$

We can quickly see that none of these answers make sense. Even if we weren't sure of the right equation, solving all of them shows that only one answer is logical.

If her weekly allowance is changed to $18, her new equation would be $18 \times w = 198$.

If we divide both sides by 18,

$$198 \div 18 = 11$$

11 weeks

17. **Part A**

(1, 6) (2, 12) (3, 18)

Part B

Newton's Second Law of Motion for a 6 lb. Object

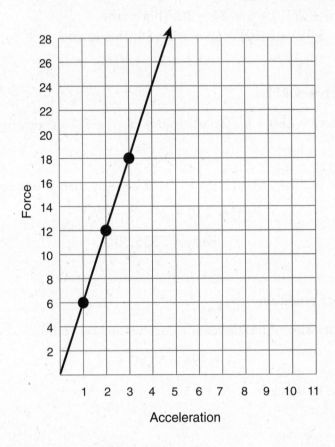

Part C

The acceleration is the independent variable because it can be any number. The force is the dependent variable, as it moves in relation to the independent variable. All relationships are shown on the line.

Just as in Chapter 8, this chapter is a practice test that lines up with testing concepts. It will be all Type 1 questions, will encompass all of the standards of sixth grade math, and will have both calculator and noncalculator sections.

This first section of approximately 20 questions is Noncalculator Active. On the real assessment remember that you must be certain of your answers in this section before moving on to the calculator section, as you will be unable to go back.

> **REMINDER:** Changes to the format, style, and questions are ongoing. Continue to check the PARCC website for up-to-date changes and information.

Practice Test 2 (No Calculator)

1. $27.6 \div 4.8$

2. Zoe has a completely fenced in rectangular garden in her backyard. The garden has a total area of $3\frac{1}{3}$ square yards. The fencing along the width is $1\frac{1}{4}$ yards long. How long is the fencing along the length?

3. Which of the following expressions are equivalent to $2(y - 5)$? Select *all* that apply.
   - ☐ A. $2y - 10$
   - ☐ B. $4(y - 5) - 2(y - 5)$
   - ☐ C. $2y - 5$
   - ☐ D. $y - 10$
   - ☐ E. $y + y + 5 + 5$
   - ☐ F. $(y - 5) + (y - 5)$

4. **Part A**

   A shoe store is having a huge blowout sale on sneakers that sell for m dollars. All sneakers are $\frac{1}{3}$ off the original price. Which expression shows the amount of money, in dollars, that Amanda saved?

   ○ A. $m + 3$

   ○ B. $m - 3$

   ○ C. $m \div 3$

   ○ D. $m \times 3$

   Part B

   If the shoes were originally $54, how much will Amanda pay for them?

5. What is the greatest common factor of 24 and 72?

6. Graph $\frac{3}{4}$ and its opposite on the number line below. What is the distance between the two points?

   <-->

7. The side of a cube measures 2.7 inches. What is the volume of the cube?

8. Colton is trying to meet a goal of swimming 100 miles this week. He has already completed 63 miles. Which inequality best represents m, the miles he still needs to swim?

   ○ A. $m > 100$

   ○ B. $m > 100 - 63$

   ○ C. $m \geq 100 - 63$

   ○ D. $m \leq 63$

9. Jon is running a fundraiser for his school's basketball team. They need to buy a new 15 foot scoreboard that costs $2,700. A local restaurant is donating $\frac{1}{5}$ of all sales on next Tuesday night to the team. The restaurant made $4,188. How much more does the team need to raise?

10. $372.36 - 12.873$

11. This coordinate plane shows the location of point V. What is the location, in decimal form, of point V?

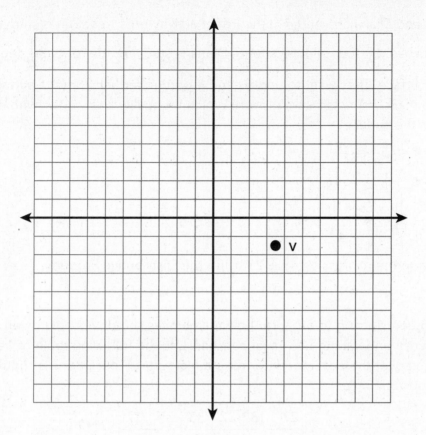

12. Mary and her sister Kate make $14\frac{1}{2}$ cups of cake batter. They put $\frac{1}{4}$ of a cup into each cupcake liner. How many cupcakes are the girls making?

13. Gabe is running in a local race. He needs to run a mile in 6.25 minutes to qualify for the top group. He currently runs a mile in 7.1 minutes. By how much does he need to reduce his time?

14. Aidan loves building with blocks. He constructs a solid right rectangular prism that is $2\frac{1}{2}$ feet wide, 3 feet long, and $1\frac{3}{4}$ feet high. How many $\frac{1}{4}$ foot cubes did he use in the construction?

15. Three sisters are decorating their room. They want to hang sparkly star lights from their ceiling. Danielle has $15.62, Kayleigh has $27.15, and Lily has $11.43. Each star costs $8.25. If they combine their money to buy as many stars as possible, what is their total cost?

16. Erin and Diana are studying the insect activity on a tree. Erin is studying the branches $4\frac{1}{2}$ feet off the ground. Diana is studying the roots $5\frac{1}{4}$ feet below the surface. They want to understand if getting farther from the surface, their zero mark, impacts insect growth. Which inequality best shows who is farther from the surface?

O A. $-5\frac{1}{4} < 4\frac{1}{2}$

O B. $-\left|-5\frac{1}{4}\right| < \left|4\frac{1}{2}\right|$

O C. $\left|-5\frac{1}{4}\right| > \left|4\frac{1}{2}\right|$

O D. $\left|-5\frac{1}{4}\right| > -4\frac{1}{2}$

17. Rewrite the expression $49 + 21$ using the Distributive Property.

_____ (_____ + _____)

18. Maureen is trying to organize how many miles she drives on a given day in order to understand her travel patterns. Use the following table of her miles driven over the past month to make a histogram. Your intervals should be in groups of 10.

27	18	22	33	35	17
44	19	31	30	8	27
25	41	14	64	50	34
20	39	30	27	22	7
42	11	36	9	23	61

19. List the following terms in order from least to greatest: $-2\frac{1}{2}$, $\left|2\frac{1}{2}\right|$, $2\frac{1}{4}$, $\left|-2\frac{3}{4}\right|$, $-\left|2\frac{1}{4}\right|$.

20. Lisa and Stephen run a candle shop. Their current stock is 4,117 candles, and they have 36 empty shelves. They want every shelf to have the exact same number of candles, with any left over to be placed at the register. How many go on each shelf? How many are used to decorate the register area?

The second section of the test begins on this page. It is calculator active. Remember you will only have access to a four function calculator (this is not a scientific calculator, meaning it will not follow order of operations for you). There are 14 questions in this section.

Note: Starting here you may use a calculator.

1. Two triangles are shown here. How much more area does the larger triangle have?

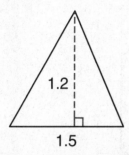
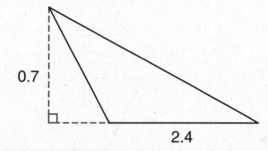

2. What is the value of $\left(\dfrac{2}{3}\right)^3$?

3. A shipping container is shown here. What is the volume in cubic yards?

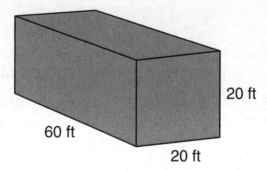

60 ft

20 ft

20 ft

4. The mean height of the 15 dancers in Addison's class is 51.2 inches. What is the total height of all the girls in the class?

5. Erica is wrapping a gift for Cara's birthday. She is trying to cover the gift exactly. What is the least amount of wrapping paper she can use on the box shown here?

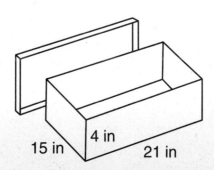

15 in   4 in   21 in

6. Blair and Grace are visiting their cousins in California, 2,454 miles away. Their trip map tells them they are 82% of the way there. How far have they traveled?

7. What is the area of this figure?

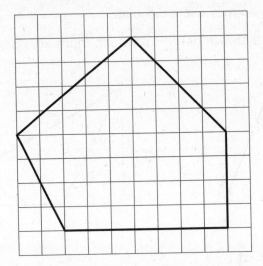

8. Hannah and Kiera have been busy selling Girl Scout cookies. This data set tells us how many cookies they sold each day.

{25, 17, 19, 23, 22, 19, 22}

They want to know their mean absolute deviation (MAD) to figure out how many cookies they might sell later this week. Find the MAD for them.

9. Rich is saving up to buy a new set of drums. The set he is looking at costs $425.85. He has already saved $89.62. He is planning to save $52.50 a week. Write an equation for the number of weeks (w) he will need to save to buy the drum set. Solve your equation.

10. Riley's rectangular dog enclosure is drawn on a grid. They need a support post at each of the corners. If a post is already located at (−3, 7), (6, −2), and (−3, −2), where do they need to put the fourth post?

11. Stephanie is making homemade ice cream. The recipe calls for 0.5 of a liter of heavy cream. How many cups is this?

12. The flag of Tom's local camp site has an orange parallelogram in the middle of a purple flag. The flag is shown here. What is the area of the visible purple?

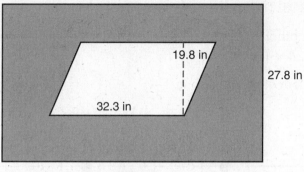

13. The double line graph below shows the number of words Karen can type per minute. How many words can she type in 13 minutes?

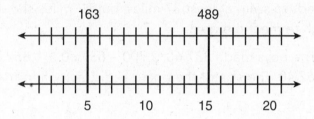

14. Solve the following with $x = 3.7$:

$$2.7 + 3x$$

$$5.92 \div x + 2.4$$

Answers

Noncalculator Portion

1. **5.75** Be sure that you move the decimal to the right out of the divisor so you are dividing 276 by 48. Use a decimal to show the remainder.

2. This requires us to divide. First, we must make both fractions improper fractions $\dfrac{10}{3} \div \dfrac{5}{4}$. Next we Keep, Change, and Flip!
$$\frac{10}{3} \times \frac{4}{5} = \frac{40}{15} = 2\frac{10}{15} = 2\frac{2}{3} \text{ yards}$$

3. **A, B**, and **F** are all correct. C did not distribute the 2 to the 5, and D did not distribute the 2 to the y. E is incorrect because it should be $y + y - 5 - 5$.

4. **Part A**

    C: $m \div 3$

Part B

54 ÷ 3 = 18. The savings is $18. To find how much Amanda must pay, we subtract 54 − 18. The shoes now cost $36.

5. 24. There are other common factors but 24 is the largest. Remember 24 is a factor of 24 (24 × 1 = 24).

6. The distance between the two is $1\frac{1}{2}$.

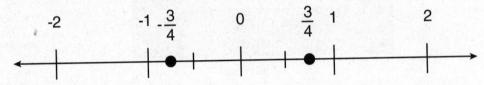

7. A cube is the same all the way around so 2.7 × 2.7 × 2.7 = 19.683 cubic inches.

8. *C*. He still needs to swim at least 37 miles, but 37 miles should be included as being enough to meet the goal.

9. 4,188 ÷ 5 = The boys made 837.60. 2,700 − 837.60 = 1,862.4. They still need to raise $1,862.40. The size of the board is distractor information.

10. 359.487

11. (3.5, −1.5)

12. $14\frac{1}{2} \div \frac{1}{4} = 14\frac{1}{2} \times 4 = 58$ cupcakes. Don't forget that when we divide fractions, we Keep, Change, Flip!

13. 7.1 − 6.25 = 0.85. He needs to reduce his time by 0.85 minute.

14. We need to find how many $\frac{1}{4}$s are in each dimension so we divide the dimensions by $\frac{1}{4}$: $2\frac{1}{2}$ feet is $10\frac{1}{4}$-inch cubes, 3 feet is $12\frac{1}{4}$-inch cubes, and $1\frac{3}{4}$ feet is $7\frac{1}{4}$-inch cubes. We find the total by multiplying the number of cubes in each dimension.

$$10 \times 12 \times 7 = 840 \text{ cubes}$$

15. Together they have a total of $54.20. It is enough to buy 6 stars for a total of $49.50 (7 stars is $57.75, which is too much!).

16. C. Distance is a measure of absolute value, and Diana is farther from the surface.

$$5\frac{1}{4} > 4\frac{1}{2}$$

17. $7(7 + 3)$

18.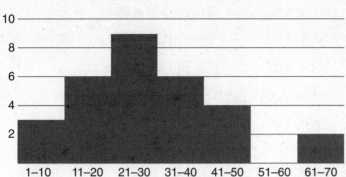

19. Remember that absolute value is only positive, so we have $-2\frac{1}{2}$, $-\left|2\frac{1}{4}\right|$, $2\frac{1}{4}$, $\left|2\frac{1}{2}\right|$, $\left|-2\frac{3}{4}\right|$.

20. There will be 114 candles per shelf, and 13 candles will be used to decorate the register area.

Calculator Portion

1. Triangle A: $\frac{1}{2}(1.2) \times 1.5 = 0.6 \times 1.5 = 0.9$ square units

   Triangle B: $\frac{1}{2}(2.4) \times 0.7 = 1.2 \times 0.7 = 0.84$ square units

   Triangle A has a bigger area by 0.06 square units.

   The slant height values are distractor information and not needed for this problem.

2. $\frac{2}{3} \times \frac{2}{3} \times \frac{2}{3} = \frac{8}{27}$

3. First, we need to change feet to yards, by dividing each by 3. The new dimensions of the container are $6.66\,(\text{or } 6\frac{2}{3}) \times 6.66\,(\text{or } 6\frac{2}{3}) \times 20 = 888.889$ or $888\frac{8}{9}$ cubic yards.

4. The average is taken by a total and divided by number of data points. So by taking the mean and multiplying it by 15, we get $51.2 \times 15 = 768$ total inches.

5. There are 3 different squares that occur twice each. 4×15, 15×21, and 21×4.

    $$4 \times 15 = 60 \times 2 = 120 \text{ square inches}$$

    $$15 \times 21 = 315 \times 2 = 630 \text{ square inches}$$

    $$21 \times 4 = 84 \times 2 = 168 \text{ square inches}$$

    $$120 + 168 + 630 = 918 \text{ square inches total}$$

6. To find percent, we multiply by the decimal $0.82 \times 2,454 = 2,012.28$. They have traveled 2,012.28 miles.

7. You need to break it into pieces. For example,

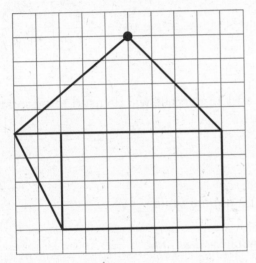

Triangle 1	$9 \times 4 \div 2 = 18$	
Triangle 2	$4 \times 2 \div 2 = 4$	
Rectangle	4×7	$= 28$
Total		$= 50$ sq units

8. First, we find the mean by adding the total boxes, which is 147. We divide this by 7 days for an average of 21 boxes. Now we find the difference between each daily sales of cookies and 21.

    4, 4, 2, 2, 1, 2, 1 (remember no negatives!)

    Now we find the average of these numbers by adding them $4 + 4 + 2 + 2 + 1 + 2 + 1 = 16$ and divide by 7, which is 2.29 (rounded to the nearest hundredth). They can expect to sell 21 boxes increased or decreased by 2.29.

9. $89.62 + 52.50w = 425.85$. He still needs to save $336.23. We divide by $52.50 per week and get 6.4. He needs to save for 7 weeks.

10. A rectangle runs along straight lines. Each side must meet at two posts. −3 is one x and is already used twice, the other side is 6. The y sides run along −2, which has two posts and 7. The missing point is (6, 7). You can solve this with logic or you could draw a sketch to help you.

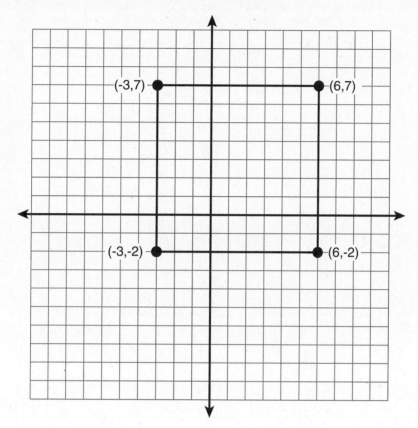

11. The conversions that we have available to us on the reference chart in Appendix A, called exhibits on the test, does not allow for us to make this conversion directly. We need to convert to gallons.

   0.5 liters = 0.132 gallons × 4 = 0.528 quarts × 2 = 1.056 pints × 2 = 2.112 cups

12. The rectangle is 47.5 inches by 27.8 inches = 1,320.5 square inches. The parallelogram has a base of 32.3 inches and a height of 19.8 = 639.54 square inches. We subtract the parallelogram from the rectangle and get

   1,320.5 − 639.54 = 680.96 square inches of purple

13. The table shows that Karen types 163 words in 5 minutes, and 489 words in 15 minutes. This is 32.6 words per minute. Multiply by 13 and get 423.8 words in 13 minutes.

14. 2.7 + (3 × 3.7) = 2.7 + 11.1 = 13.8

   5.92 ÷ 3.7 + 2.4 = 1.6 + 2.4 = 4

Rates, Ratios, and Proportions

CCSS.MATH.CONTENT.6.RP.A.1

Understand the concept of a ratio and use ratio language to describe a ratio relationship between two quantities.

CCSS.MATH.CONTENT.6.RP.A.2

Understand the concept of a unit rate a/b associated with a ratio $a:b$ with $b \neq 0$, and use rate language in the context of a ratio relationship.

CCSS.MATH.CONTENT.6.RP.A.3

Use ratio and rate reasoning to solve real-world and mathematical problems, e.g., by reasoning about tables of equivalent ratios, tape diagrams, double number line diagrams, or equations.

CCSS.MATH.CONTENT.6.RP.A.3.A

Make tables of equivalent ratios relating quantities with whole-number measurements, find missing values in the tables, and plot the pairs of values on the coordinate plane. Use tables to compare ratios.

CCSS.MATH.CONTENT.6.RP.A.3.B

Solve unit rate problems including those involving unit pricing and constant speed.

CCSS.MATH.CONTENT.6.RP.A.3.C

Find a percent of a quantity as a rate per 100 (e.g., 30% of a quantity means 30/100 times the quantity); solve problems involving finding the whole, given a part and the percent.

CCSS.MATH.CONTENT.6.RP.A.3.D

Use ratio reasoning to convert measurement units; manipulate and transform units appropriately when multiplying or dividing quantities.

Number System

Apply and extend previous understandings of multiplication and division to divide fractions by fractions.

CCSS.MATH.CONTENT.6.NS.A.1

Interpret and compute quotients of fractions, and solve word problems involving division of fractions by fractions, e.g., by using visual fraction models and equations to represent the problem.

Compute fluently with multi-digit numbers and find common factors and multiples.

CCSS.MATH.CONTENT.6.NS.B.2

Fluently divide multi-digit numbers using the standard algorithm.

CCSS.MATH.CONTENT.6.NS.B.3

Fluently add, subtract, multiply, and divide multi-digit decimals using the standard algorithm for each operation.

CCSS.MATH.CONTENT.6.NS.B.4

Find the greatest common factor of two whole numbers less than or equal to 100 and the least common multiple of two whole numbers less than or equal to 12. Use the distributive property to express a sum of two whole numbers 1–100 with a common factor as a multiple of a sum of two whole numbers with no common factor.

Apply and extend previous understandings of numbers to the system of rational numbers.

CCSS.MATH.CONTENT.6.NS.C.5

Understand that positive and negative numbers are used together to describe quantities having opposite directions or values (e.g., temperature above/below zero, elevation above/below sea level, credits/debits, positive/negative electric charge); use positive and negative numbers to represent quantities in real-world contexts, explaining the meaning of 0 in each situation.

CCSS.MATH.CONTENT.6.NS.C.6

Understand a rational number as a point on the number line. Extend number line diagrams and coordinate axes familiar from previous grades to represent points on the line and in the plane with negative number coordinates.

CCSS.MATH.CONTENT.6.NS.C.6.A

Recognize opposite signs of numbers as indicating locations on opposite sides of 0 on the number line; recognize that the opposite of the opposite of a number is the number itself, e.g., –(–3) = 3, and that 0 is its own opposite.

CCSS.MATH.CONTENT.6.NS.C.6.B

Understand signs of numbers in ordered pairs as indicating locations in quadrants of the coordinate plane; recognize that when two ordered pairs differ only by signs, the locations of the points are related by reflections across one or both axes.

CCSS.MATH.CONTENT.6.NS.C.6.C

Find and position integers and other rational numbers on a horizontal or vertical number line diagram; find and position pairs of integers and other rational numbers on a coordinate plane.

CCSS.MATH.CONTENT.6.NS.C.7

Understand ordering and absolute value of rational numbers.

CCSS.MATH.CONTENT.6.NS.C.7.A

Interpret statements of inequality as statements about the relative position of two numbers on a number line diagram.

CCSS.MATH.CONTENT.6.NS.C.7.B

Write, interpret, and explain statements of order for rational numbers in real-world contexts.

CCSS.MATH.CONTENT.6.NS.C.7.C

Understand the absolute value of a rational number as its distance from 0 on the number line; interpret absolute value as magnitude for a positive or negative quantity in a real-world situation.

CCSS.MATH.CONTENT.6.NS.C.7.D

Distinguish comparisons of absolute value from statements about order.

CCSS.MATH.CONTENT.6.NS.C.8

Solve real-world and mathematical problems by graphing points in all four quadrants of the coordinate plane. Include use of coordinates and absolute value to find distances between points with the same first coordinate or the same second coordinate.

Expressions and Equations

Apply and extend previous understandings of arithmetic to algebraic expressions.

CCSS.MATH.CONTENT.6.EE.A.1

Write and evaluate numerical expressions involving whole-number exponents.

CCSS.MATH.CONTENT.6.EE.A.2

Write, read, and evaluate expressions in which letters stand for numbers.

CCSS.MATH.CONTENT.6.EE.A.2.A

Write expressions that record operations with numbers and with letters standing for numbers.

CCSS.MATH.CONTENT.6.EE.A.2.B

Identify parts of an expression using mathematical terms (sum, term, product, factor, quotient, coefficient); view one or more parts of an expression as a single entity.

CCSS.MATH.CONTENT.6.EE.A.2.C

Evaluate expressions at specific values of their variables. Include expressions that arise from formulas used in real-world problems. Perform arithmetic operations, including those involving whole-number exponents, in the conventional order when there are no parentheses to specify a particular order (Order of Operations).

CCSS.MATH.CONTENT.6.EE.A.3

Apply the properties of operations to generate equivalent expressions.

CCSS.MATH.CONTENT.6.EE.A.4

Identify when two expressions are equivalent (i.e., when the two expressions name the same number regardless of which value is substituted into them).

Reason about and solve one-variable equations and inequalities.

CCSS.MATH.CONTENT.6.EE.B.5

Understand solving an equation or inequality as a process of answering a question: Which values from a specified set, if any, make the equation or inequality true? Use substitution to determine whether a given number in a specified set makes an equation or inequality true.

CCSS.MATH.CONTENT.6.EE.B.6

Use variables to represent numbers and write expressions when solving a real-world or mathematical problem; understand that a variable can represent an unknown number, or, depending on the purpose at hand, any number in a specified set.

CCSS.MATH.CONTENT.6.EE.B.7

Solve real-world and mathematical problems by writing and solving equations of the form $x + p = q$ and $px = q$ for cases in which p, q, and x are all nonnegative rational numbers.

CCSS.MATH.CONTENT.6.EE.B.8

Write an inequality of the form $x > c$ or $x < c$ to represent a constraint or condition in a real-world or mathematical problem. Recognize that inequalities of the form $x > c$ or $x < c$ have infinitely many solutions; represent solutions of such inequalities on number line diagrams.

Represent and analyze quantitative relationships between dependent and independent variables.

CCSS.MATH.CONTENT.6.EE.C.9

Use variables to represent two quantities in a real-world problem that change in relationship to one another; write an equation to express one quantity, thought of as the dependent variable, in terms of the other quantity, thought of as the independent variable. Analyze the relationship between the dependent and independent variables using graphs and tables, and relate these to the equation. For example, in a problem involving motion at constant speed, list and graph ordered pairs of distances and times, and write the equation $d = 65t$ to represent the relationship between distance and time.

Geometry

Solve real-world and mathematical problems involving area, surface area, and volume.

CCSS.MATH.CONTENT.6.G.A.1

Find the area of right triangles, other triangles, special quadrilaterals, and polygons by composing into rectangles or decomposing into triangles and other shapes; apply these techniques in the context of solving real-world and mathematical problems.

CCSS.MATH.CONTENT.6.G.A.2

Find the volume of a right rectangular prism with fractional edge lengths by packing it with unit cubes of the appropriate unit fraction edge lengths, and show that the volume is the same as would be found by multiplying the edge lengths of the prism. Apply the formulas $V = lwh$ and $V = bh$ to find volumes of right rectangular prisms with fractional edge lengths in the context of solving real-world and mathematical problems.

CCSS.MATH.CONTENT.6.G.A.3

Draw polygons in the coordinate plane given coordinates for the vertices; use coordinates to find the length of a side joining points with the same first coordinate or the same second coordinate. Apply these techniques in the context of solving real-world and mathematical problems.

CCSS.MATH.CONTENT.6.G.A.4

Represent three-dimensional figures using nets made up of rectangles and triangles, and use the nets to find the surface area of these figures. Apply these techniques in the context of solving real-world and mathematical problems.

Statistics and Probability

Develop understanding of statistical variability.

CCSS.MATH.CONTENT.6.SP.A.1

Recognize a statistical question as one that anticipates variability in the data related to the question and accounts for it in the answers.

CCSS.MATH.CONTENT.6.SP.A.2

Understand that a set of data collected to answer a statistical question has a distribution which can be described by its center, spread, and overall shape.

CCSS.MATH.CONTENT.6.SP.A.3

Recognize that a measure of center for a numerical data set summarizes all of its values with a single number, while a measure of variation describes how its values vary with a single number.

Summarize and describe distributions.

CCSS.MATH.CONTENT.6.SP.B.4

Display numerical data in plots on a number line, including dot plots, histograms, and box plots.

CCSS.MATH.CONTENT.6.SP.B.5

Summarize numerical data sets in relation to their context, such as by:

CCSS.MATH.CONTENT.6.SP.B.5.A

Reporting the number of observations.

CCSS.MATH.CONTENT.6.SP.B.5.B

Describing the nature of the attribute under investigation, including how it was measured and its units of measurement.

CCSS.MATH.CONTENT.6.SP.B.5.C

Giving quantitative measures of center (median and/or mean) and variability (interquartile range and/or mean absolute deviation), as well as describing any overall pattern and any striking deviations from the overall pattern with reference to the context in which the data were gathered.

CCSS.MATH.CONTENT.6.SP.B.5.D

Relating the choice of measures of center and variability to the shape of the data distribution and the context in which the data were gathered.

Standards for Mathematical Practices

CCSS.MATH.PRACTICE.MP1 Make sense of problems and persevere in solving them.

Mathematically proficient students start by explaining to themselves the meaning of a problem and looking for entry points to its solution. They look for key words to understand what is being asked as well as for various ways to solve and demonstrate understanding. They know when a chosen method is not working and look for an alternative.

CCSS.MATH.PRACTICE.MP2 Reason abstractly and quantitatively.

Mathematically proficient students make sense of quantities and their relationships in problem situations. They are able to take mathematical rules and apply them to real-world situations and to take real-life scenarios and turn them into mathematical understanding. They are able to move from the concrete to mathematical theory and back again.

CCSS.MATH.PRACTICE.MP3 Construct viable arguments and critique the reasoning of others.

Mathematically proficient students accurately judge and critique mathematical thought. They can defend their own judgment and are able to explain the fault in mathematical errors. Not only can they determine that an answer is incorrect, but they are able to fix it, furthering their own mathematical thought. When an error is found in their work, they are able to absorb the corrections to improve their understanding.

CCSS.MATH.PRACTICE.MP4 Model with mathematics.

Mathematically proficient students can apply the mathematics they know to solve problems arising in everyday life, society, and the workplace. Using equations, drawings, diagrams, function tables, charts and graphs, they can demonstrate mathematical relationships and express mathematical thinking.

CCSS.MATH.PRACTICE.MP5 Use appropriate tools strategically.

Mathematically proficient students consider the available tools when solving a mathematical problem. They understand how best to use physical tools such as rulers, but they also know how to use computer software, models, and formulas. They recognize the tools that need to be used and use them correctly. They check their work for accuracy.

CCSS.MATH.PRACTICE.MP6 Attend to precision.

Mathematically proficient students try to communicate precisely to others. They carefully select terms, vocabulary, operations, and numbers to create solutions in the clearest way. They work with high levels of accuracy.

CCSS.MATH.PRACTICE.MP7 Look for and make use of structure.

Mathematically proficient students look closely to discern a pattern or structure. They break down numbers into factors and reorder terms to come up with equivalent terms. They can take a step back from a problem and come at it from a different direction because they noticed another way may be easier.

CCSS.MATH.PRACTICE.MP8 Look for and express regularity in repeated reasoning.

They look for repeated calculations, identifying things such as repeating decimals. They look for patterns when computing to derive formulas.

Mathematics Grade 6 Score Report

About the Score Report

After the PARCC has been completed, a report will be generated in order for you and your parents to understand your level on the test. Score reports for the first year were generated in the following fall. Moving forward PARCC is hoping to have all score reports to you before the start of summer. Score reports similar to this will also be provided to your school and your teachers so that they can help you do even better in school.

When you receive your score report it will contain a key to help you understand the information on the report. The report is designed in three parts. Section one is your Performance Level from 1–5, which corresponds to how you met the standard. The second is a scaled 3-digit score that can be used to compare your score to other students that took the test. Here you can see how you match up with your school, your district, your state, and all other 6th graders who took the same test. Section three is a detailed account of strengths and weaknesses you showed in the sections. This can tell you what to focus on for the next year.

To view a mock student report, go to the PARCC website:

http://www.parcconline.org/assessments/score-results

Report for Parents

1. **What are the PARCC assessments?** The PARCC (Partnership for Assessment of Readiness for College and Careers) assessments are designed to measure and report the extent to which students in grades 3–11 have learned the grade-level material in English language arts/literacy and mathematics that will prepare them for the next grade level and eventually for college and careers.

2. **How can I use this report to help my child?** Use the report as a springboard for discussion with your child's teacher(s) about his/her academic strengths and areas for improvement.

3. **What do the scores on the report mean?**

- **Performance Levels.** Your child's overall score falls into one of five performance levels. The levels are an indication to the extent to which your child has demonstrated grade-level material. Students achieving levels 4 and 5 have demonstrated a strong grasp of grade-level material and are well-prepared for the next grade level. See examples of the kinds of test questions that students at each performance level can typically answer at *http://www.parcconline.org/examples*.

- **Mathematics Score.** As each performance level contains a range of scores, this shows where within the performance level your child scored.

- **Additional information about your child's mathematics score.** This section provides information about your child's strengths and areas for improvement. In each area, you can see how your child did compared to students who performed at Level 4 overall in Mathematics.

- **Margin of error.** The amount of change that would be expected in your child's score if he/she were to take the test many times. Learn more about this report and what it means at *www.parcconline.org*.

Performance Level Definitions

Below is a brief description of how well students demonstrate understanding of subject matter at each performance level, along with an indication of their academic preparedness for further studies in Mathematics:

- **Level 5:** Student demonstrated a **distinguished understanding** of subject matter and is academically well prepared to engage successfully in further studies.

- **Level 4:** Student demonstrated a **strong understanding** of subject matter and is academically prepared to engage successfully in further studies.

- **Level 3:** Student demonstrated a **moderate understanding** of subject matter and will likely need academic support to engage successfully in further studies.

- **Level 2:** Student demonstrated a **partial understanding** of subject matter and will need academic support to engage successfully in further studies.

- **Level 1:** Student demonstrated a **minimal understanding** of subject matter and will need extensive academic support to engage successfully in further studies.

- Students in grade 6 will be provided a reference sheet with the information shown below. Notice that the names of the measurement formulas provided on the reference sheet include only the name of the figure or object to which the measurement formula(s) is applied. The intent of the Common Core State Standards in Mathematics at grade 6 is to know and apply the measurement formulas. In order for students to be able to choose the correct formula, they will need to know the formula.

Grade 6

Reference Sheet

1 inch = 2.54 centimeters	1 kilometer = 0.62 mile	1 cup = 8 fluid ounces
1 meter = 39.37 inches	1 pound = 16 ounces	1 pint = 2 cups
1 mile = 5280 feet	1 pound = 0.454 kilograms	1 quart = 2 pints
1 mile = 1760 yards	1 kilogram = 2.2 pounds	1 gallon = 4 quarts
1 mile = 1.609 kilometers	1 ton = 2000 pounds	1 gallon = 3.785 liters
		1 liter = 0.264 gallons
		1 liter = 1000 cubic centimeters

Triangle	$A = \frac{1}{2}bh$
Rectangular Prism	$V = bh$ or $V = llh$

- Students in grade 6 will be required to know relative sizes of measurement units within one system of units. Therefore, the following requisite knowledge is necessary for the grade 6 assessments and is **not** provided in the reference sheet.

1 meter = 100 centimeters	1 foot = 12 inches
1 meter = 1000 millimeters	1 yard = 3 feet
1 kilometer = 1000 meters	1 day = 24 hours
1 kilogram = 1000 grams	1 minute = 60 seconds
1 liter = 1000 milliliters	1 hour = 60 minutes

The formulas for the area of a rectangle are also considered to be requisite knowledge because the intent of the Common Core State Standards in Mathematics for students in grade 6 is to have a conceptual understanding of area of rectangles.

Area of a Rectangle	$A = ll$ or $A = bh$

Grade 6 ruler provided on the PARCC paper-based mathematics assessments (not actual size):

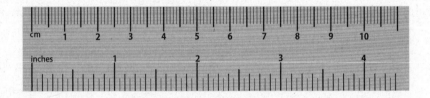

Grade 6 protractor provided on the PARCC paper-based mathematics assessments (not actual size):

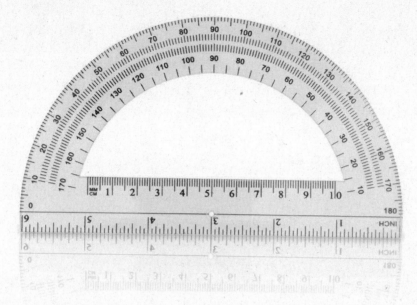

Scrap Paper (required):

- Blank scratch paper (graph, lined, or un-lined paper) is intended for use by students to take notes and work through items during testing. At least two pages per unit must be provided to each student. Any work on scratch paper will **not** be scored.

Index